10th International Workshop NDT in Progress

Prague, Czech Republic
7 – 9 October 2019

ISBN: 978-1-7138-5225-4

Printed from e-media with permission by:

Curran Associates, Inc.
57 Morehouse Lane
Red Hook, NY 12571

Some format issues inherent in the e-media version may also appear in this print version.

Printed with permission by Curran Associates, Inc. (2022)

For permission requests, please contact NDT.net
at the address below.

NDT.net
Rheintalstr 3
53498 Bad Breisig
Germany

Phone: + 49 2633 4756670
Fax: + 49 3212-1030513

info@ndt.net

Additional copies of this publication are available from:

Curran Associates, Inc.
57 Morehouse Lane
Red Hook, NY 12571 USA
Phone: 845-758-0400
Fax: 845-758-2633
Email: curran@proceedings.com
Web: www.proceedings.com

TABLE OF CONTENTS

Improvement of microwave sensing of woods using metamaterials structures

Dagmar FAKTOROVA[1], Rozina STEIGMANN[2], Nicoleta IFTIMIE[2], Gabriel Silviu DOBRESCU[2], Adriana SAVIN[2], Mariana Domnica STANCIU[3]

[1]Faculty of Electrical Engineering and Information Technology, University of Žilina, Slovak Republic
[2] Nondestructive Testing Department, National Institute of R&D for Technical Physics, Iasi, Romania
[3] Faculty of Mechanical Engineering, Transilvania University, Brasov, Romania

Abstract

The aim is to achieve the optimal design of metamaterial structure for waveguide sensor tuning in microwave X-band. The numerical simulation of 2D metamaterial structure properties and experimental results for dielectric properties dielectric materials are carried out. In a large range of applications, the information about dielectric properties are required, but not only these, sometimes, for natural dielectric as wood, the direction of the wood grain that is changing from point to point, making the permittivity tensor to become a random values. The purpose of the paper was to apply a modern microwave measurement method to determine the dielectric properties of wood samples and validate the results with Dynamic Mechanical Analysis.

Keywords: microwave testing, wood species, metamaterials, dielectric properties

1. Introduction

There are 1.43 million ha of Norway spruce (*Picea abies* L. (Krast)) in the Romanian Carpathians [1] which host one of the most important resources of resonance wood in Europe, in terms of value and volume. Although it is continuously decreasing, the resonance wood resource still satisfies the demand for the local musical instruments industry which are shipped across all continents.

Dielectric materials are the poor electric conductors (non-ideal insulator). Starting from paper and ending to special dielectrics used in solar cells, dielectrics are used in various domains with interesting practical applications. The dielectric properties determine the electrical features of materials, and these give information useful for their further utilization and characterization [2]. The dielectric properties are usually determined in resonant circuits [3], electronic bridges [4], transient methods in DC [5], etc. Measurements on low-loss materials using closed and open cavity resonators, and dielectric resonator methods have been used since 1990 [6] but microwaves (MW) measurements can be carried out now by velocity-modulated tubes, cavities, waveguide technologies and free space configurations [7]. Using the interactions between microwave and the materials, microwave waveguide sensors [8] can measure dielectric properties of materials.

Microwave waveguide sensors can measure properties of materials based on microwaves interaction with matter, and they can be used to provide information about dielectric properties of investigated dielectric material characterized with complex permittivity, and with that knowledge can afford information about moisture content, density, structure, and even chemical reaction [9]. Microwave sensor offers many advantages in comparison with conventional sensor such as rapid and nondestructive measurement [10]. Nowadays, a great interest has been devoted to the sensing applications of metamaterials to increase the sensing properties of conventional waveguide sensor. Not only absolute value of complex permittivity can be in interest – in some technological processes it is needed information about changing dielectric properties of processed materials, but, for some natural materials as wood, the direction of the wood grain is changing from point to point and makes the permittivity tensor to be an aleatory quantity [11]. Wood is a natural, renewable material, which is used in a wide variety of products such as load bearing structures and furniture. However, the heterogeneous nature of wood leads to a large variability of its physical properties, which creates serious difficulties in industrial processes where well-defined material parameters are essential for

manufacturing with consistent quality. Over the past decades the traditional use of wood as construction material evolved to a modern, widely differentiated application domains, as sonic barriers or components in small wind turbine blades, etc.
The purpose of the paper was to apply a modern microwave measurement method to determine the dielectric properties of wood samples and compare the results with Dynamic Mechanical Analysis.

2. Samples and methods

2.1. Samples

Wood is an anisotropic dielectric which, greatly simplifying the physical reality, can be considered as a triaxial crystal, with three different principle axes related to the wood grain. Modern production lines demand sophisticated methods for quality control, and the implementation of new technologies in the wood industry is promising therefore.
The studied samples were cropped from plates of spruce [12], to fit the inside of the rectangular microwave waveguide and others having the characteristics described in table 1. The density was determined with DPX300 Machine.

Table 1. Samples characteristics for DMA testing

Cases - Load direction reported to wood structure	Samples	Moisture Contents %	Density ρ[g/cm^3]	Sizes [mm] *L*	*l*	*h*
Case 1 Tangential Bending Moment (M_{bTg})	**Spruce 1.**	13.9	0.441	50	10.63	4.95
	Spruce 2.	14.2	0.423	50	10.73	4.90
Case 2 Radial Bending Moment (M_{bR})	**Spruce 3.**	13.7	0.397	50	10.70	4.90
	Spruce 4.	13.9	0.366	50	10.6	4.99

2.2. Microwave testing

Measurements were done on Vector Network Analyzer VNA MS2028C with connected close waveguide section with metamaterial structure in front of waveguide opening (Figure 1).

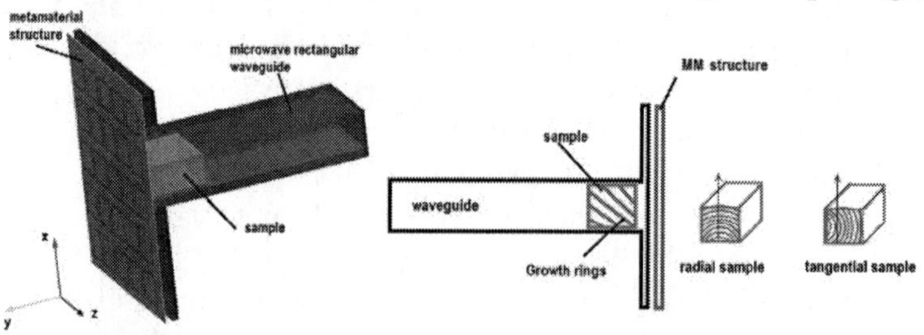

Figure 1. Microwave testing experimental set-up

The samples – Spruce wood – were placed in waveguide hard by the walls of waveguide, therefore the corrections were not necessary in relative permittivity calculation. The transmission – reflection method was used for measurement and Nicolson-Ross-Weir (NRW) conversion process [13,14] was used for relative permittivity calculation. The values of S_{11} and S_{21} parameters, which were measured and parameters of critical wavelength in waveguide, the sample length and the wavelength in free space were used in calculation. Thus, transmission (T) and reflection (Γ) coefficients are extracted from S-parameters measured by VNA using the scheme presented in Figure 2.

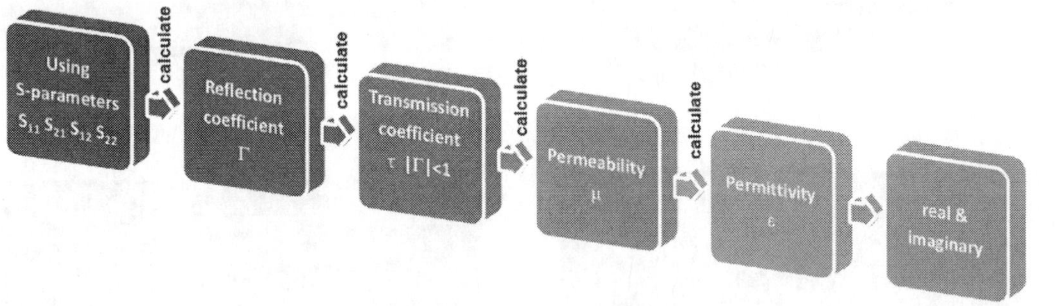

Figure 2. Schematic procedure of NRW method

The reflection coefficient Γ is calculated [15]

$\Gamma = X \mp \sqrt{X^2 - 1}$ where $|\Gamma| < 1$, in order to calculate the root, X being replaced consequently by S parameters. $X = \dfrac{S_{11}^2 - S_{21}^2 + 1}{2S_{11}}$. The transmission coefficient is calculated $T = \dfrac{S_{11} + S_{21} - \Gamma}{1 - \Gamma(S_{11} + S_{21})}$, noting Λ as $\dfrac{1}{\Lambda^2} = \left(\dfrac{j}{2\pi d}\ln(T)\right)^2$. Using cut-off wavelength λ_c and free space wavelength λ_0, λ_{0g} can be determined $\lambda_{0g} = \dfrac{1}{\sqrt{\dfrac{1}{\lambda_0^2} - \dfrac{1}{\lambda_c^2}}}$. Thus, complex permeability and respective permittivity are given by

$$\mu^* = \frac{\lambda_{0g}}{\Lambda}\left(\frac{1+\Gamma}{1-\Gamma}\right) \text{ and } \varepsilon^* = \frac{\lambda_0^2\left(\dfrac{1}{\Lambda^2} + \dfrac{1}{\lambda_c^2}\right)}{\mu^*}.$$

Thus, only transmission S_{11} and reflection S_{21} are used to extract the complex permittivity.

2.3. DMA testing

The three points bending flexural test under force was performed. Bending moment (Mb) is orthogonal to force direction as can be seen in Fig. 3a. Two cases of load were considered in accordance with type of flexural and disposal of bending moment to wood fibers. The storage modulus E', loss modulus E" and damping *tanδ* have been determined using Dynamic Mechanical Analyzer DMA 242C- Netzsch.

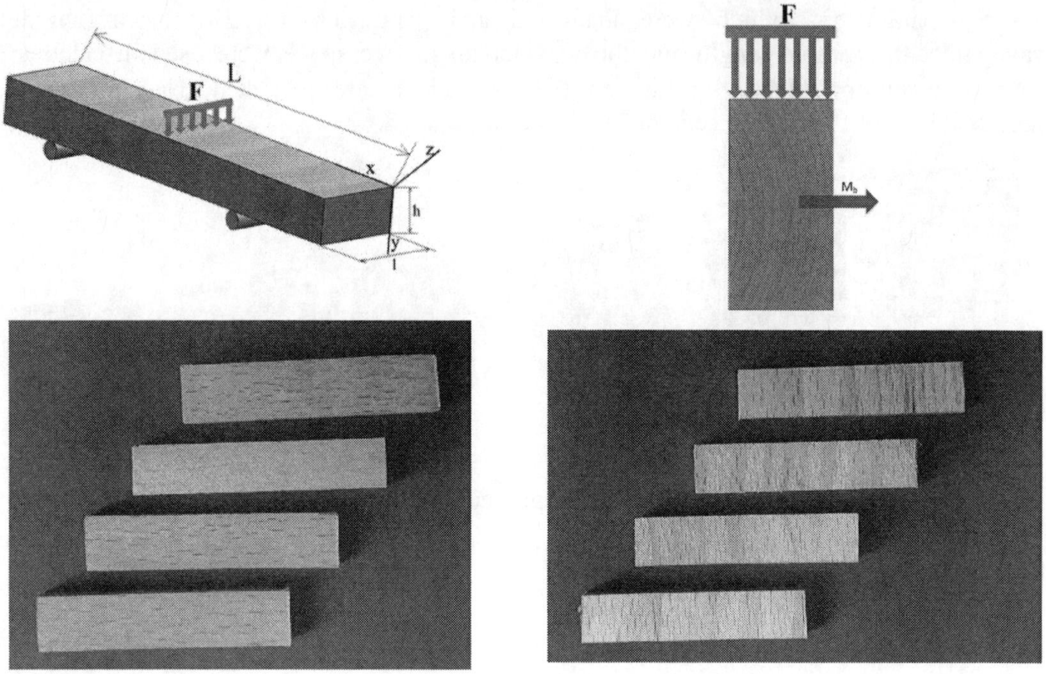

a b

Figure 3. Load direction reported to wood structure for studied samples: a – Set-up of beam for dynamic bending test according to ASTM D5023 – 07; b – case 2 (Radial Bending Moment M_{bR})

The measurement conditions were: isotherm measurement at 30 ± 0.1^0C, frequency 1Hz, during 10 minutes, the amplitude of force 6N and maximum deflection 30pm.

3. Numerical simulation

In order to obtain the distribution of the electromagnetic field, a finite difference time domain (FDTD) method is applied using the scheme suggested by [16] to the Maxwell equations. The electric field vector components are orthogonal to the magnetic field vector components upon half-cell. The electric field and magnetic field are evaluated at alternative half time steps. For TE mode, the electric and magnetic field components are expressed by the total field FDTD equations. The simulation has been carried out in XFDTD software by REMCOM [17].The simulations of the typical electric field of TE_{10} mode along the center axis of rectangular wave guide size=22.86mm × 10.16mm: a) Rectangular wave guide is empty, ε_r=1and b) Rectangular wave guide is filled with wood sample, operating at the microwave frequency of 7.5GHz (30 s). The time is set-up short because it is known that microwaving wood leads to drying and dielectric properties vary with temperature. Figure 4 shows the electric field distribution, respective the magnetic field, inside the rectangular wave guide with wood sample inside at the frequency of 7.5GHz. Since the typical depth of wood is close to the penetration depth of microwave [16], the microwaves penetrate through the sample. The wave will be reflected on each interface, from air (cavity) to top surface and from lower surface of sample to air (cavity).

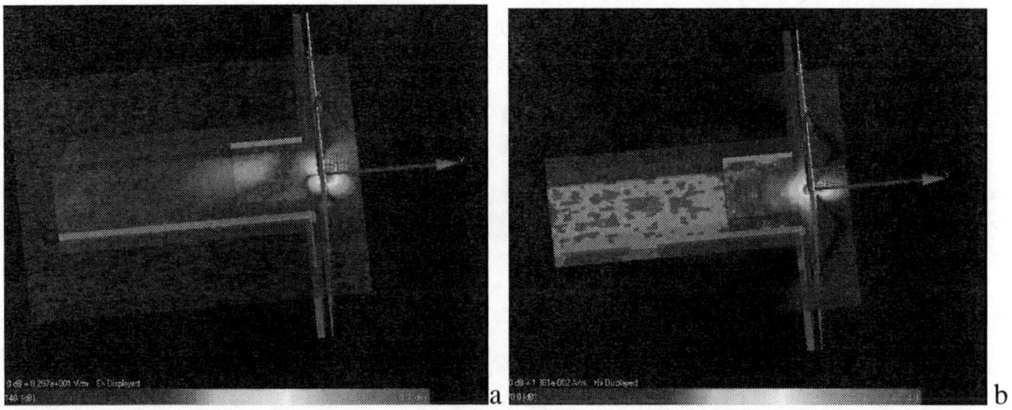

Figure 4. Distribution of electromagnetic field with sample inside: a) E_x; b) H_x

The reflection and transmission components at each interface will contribute to the resonance of stationary wave configuration inside the sample and give rise to a microwave absorption peak further from the surface exposed to incident microwaves. It can be seen that a stronger standing wave with large amplitude is formed by interference between the incident and reflected wave from the surface of sample due to the difference of dielectric properties of materials (air and sample) at this interface.

4. Experimental results

The S parameters of metamaterials structure have been measured with Network Analyzer Agilent E5071B USA18. The results are presented in Figure 5.

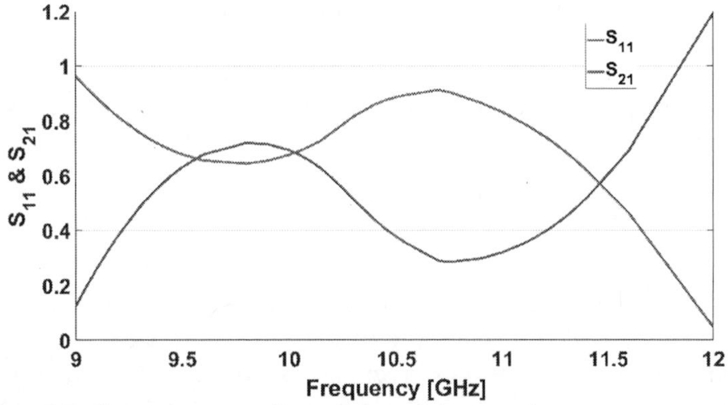

Figure 5. S parameters of metamaterial structure

The experimental measurements have been carried out using the metamaterial structure in the front of a rectangular waveguide BJ-100 with dimensions a = 22.86mm, b = 10.16mm. The length of the waveguide used is 60mm, the waveguide walls are PEC, the end of the waveguide is a metallic grounding square shape of size 40mm x 50mm. The metamaterial layer consist in 5×5 split ring resonators (SRR) whose resonant frequency corresponds to the operating frequency of the system [19]. The metamaterial structure was etched on the ROGERS RT/DUROID 5870. The dielectric sample was placed in the near electromagnetic field area of waveguide sensor tuned with 2D metamaterial structure, Figure 1. Two type of samples having same high and width to fit inside the waveguide were taken into consideration: sample 1 having 50mm length and sample 2 having 30mm length. The aim is to find the shift of resonant frequency due to changing of sample dielectric properties with frequency. The results of measurements are given in Figure 6.

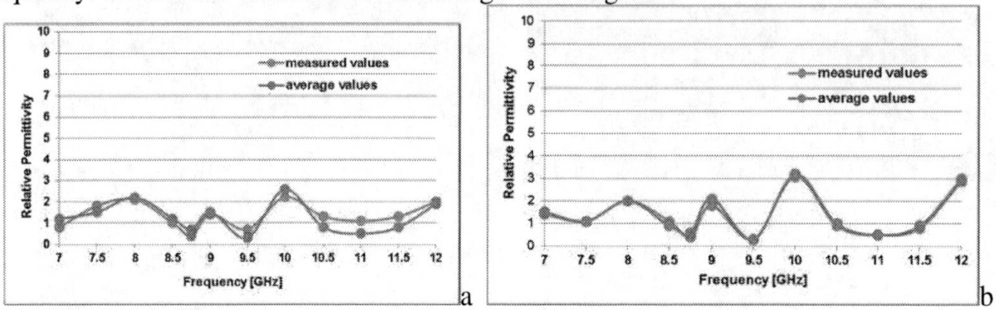

Figure 6. Microwave measurements - dielectric constant vs frequency: a) sample 1; b) sample 2

In order to validate the method, the samples were destructive tested using Dynamic Mechanical Analyzer DMA 242C by Netzsch Germany.

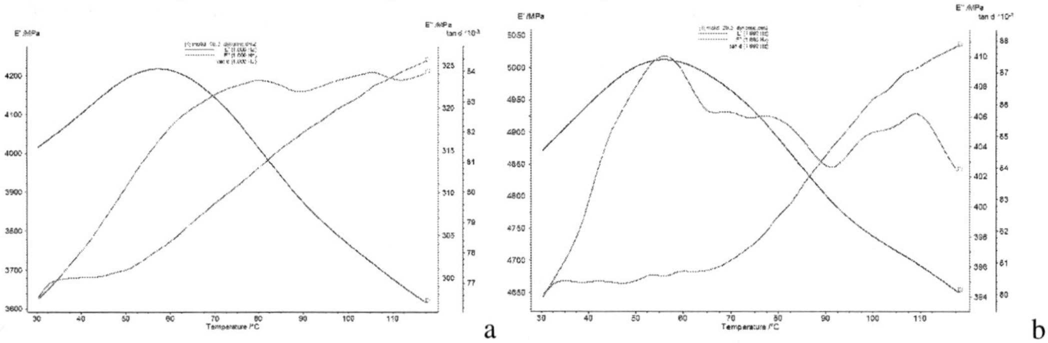

Figure 7. Correlation between storage and loss modulus of Spruce versus time: a) radial rare fibre; b) semiradial rare fibre

Dynamical modulus of elasticity is lower than statically one regardless of loading. For example, in case of longitudinal flexural, statically elasticity modulus is 2.34 times higher than elasticity modulus determined with ultrasound method [20, 21] and 2.77 times higher than dynamic modulus with DMA.

Conclusions

In the paper we investigate the possibility of artificial metamaterial structure using in waveguide microwave method for upgrading properties of classical waveguide sensor. The microwave methods are nowadays used especially for the dielectric materials investigation, their advantages allowed us to measure without contact medium. The dielectric properties of wooden samples were measured in X – band frequency region and results showed the frequency dependence of wood dielectric properties. The knowledge of this dependence is important for very precise tracking the dielectric properties changes connecting with various factors which occur in processes connected with wood processing and is important also in restoration processes of old wood artefacts where the quality of old wood should be known.

The numerical analysis has allowed the determination of optimal frequency required to excite the microwave waveguide in order to test wood samples. The mathematical allowed the determination of reflection and transmission coefficients using S-parameters. Using the results obtained by DMA, and completing with dielectric characteristics, a complete characterization of Spruce samples taken into consideration, promoting them for use in different practical application as sound barrier, acoustic cavities, etc.

Future works will take in analysis other local species of wood for further development of new applications taking into consideration also the moisture content.

Acknowledgements

This work was partially supported by a grant of the Romanian Ministry of Research and Innovation, CCCDI – UEFISCDI, project number PN-III-P1-1.2-PCCDI-2017-0239 within PNCDI III and Nucleus Program PN 19 26 01 02 and by the Slovak Research and Development Agency under the contract No. APVV-17-0218.

References

[1] https://silvanews.ro/silvicultura/dendrologie/molid/
[2] Baker-Jarvis J, Geyer RG, Grosvenor JH, Janezic MD, Jones CA, Riddle B, Weil CM and Krupka J., 1998. IEEE Transactions on Dielectrics and Electrical insulation, 5(4): 571-577.
[3] Anis M, Jostingmeier A, Meyer T, Omar AS 2005 International Symposium on Antennas and Propagation Society, Vol. 1:421-424
[4] Singh S, Mohsin MM, Masood A 2016 IEEE International Conference on Power Electronics, Intelligent Control and Energy Systems (ICPEICES):1-4
[5] Walker E, Glover PWJ, Ruel J 2014 Journal of Geophysical Research: Solid Earth, 119(2):957-970
[6] Bahl I and Ely K 1990 Modern microwave substrate materials, Microwave J. Vol. 33:. 131-146.
[7] Okress EC ed 2017 Microwave Power Engineering: Generation, Transmission, Rectification, Vol. 1, London: Academic Press
[8] Istenikova K and Faktorova D 2012 Electrical Review 88(7b): 223-225
[9] Baena JD, Bonache J, Martin F, Sillero RM, Falcone F, Lopetegi T, Laso MA, Garcia-Garcia J, Gil I, Portillo MF and Sorolla M 2005 IEEE transactions on microwave theory and techniques,53(4): 1451-1461
[10] Lai A, Itoh T and Caloz C, 2004.IEEE microwave magazine,5(3): 34-50.
[11] Torgovnikov GI 1993 Dielectric Properties of Wood and Wood-Based Materials. Springer Verlag, Berlin.
[12] Stanciu, M.D., Curtu, I., Grimberg, R. and Savin, A., 2013. Pro Ligno, 9(4), pp.587-593.
[13] Nicolson A, Ross G 1970 IEEE Transactions on Instrumentation and Measurement,19: 377-382.
[14] Weir WB 1974 Proceedings of the IEEE. 1974;62:33-36.
[15] Rutpralom T, Chamnongthai K, Kumhom P, Krairiksh M 2006 IEEE International Symposium on Circuits & Systems: 1351-1354.
[16] Rattanadecho P, 2006 Chemical Engineering Science 61(14): 4798-4811.
[17] ZhangQ, Yuan CW, Liu L 2012 Theoretical Design and Analysis for TE20-TE10 Rectangular Waveguide Mode Converters. IEEE Transactions on Microwave Theory and Techniques 60: 1018-1026.
[18] Savin, A, Steigmann R, Bruma A, Sturm R, 2015 Sensors, 15(7): 15903-15920
[19] Iftimie N, Faktorova D, Fabo P, Savin A, Steigmann R, 2018 IOP Conference Series: Materials Science and Engineering, Vol. 444, No. 2, 022007
[20] Bucur V 2006 Acoustics of Wood. Springer, Berlin
[21] Pánek, M. and Trgala, K., 2016. Forestry Journal, 62(3), pp.164-168.

Magnetic NDE system for detection of crack in steel deck with high lift-off using secondary differential detection coil

Yuto Goda [1] Hiroto Shobu [1] Takumi Kobara [1] Kenji Sakai [1] Toshihiko Kiwa [1]
Akira Tsukamoto [2] Keiichi Tanabe [2] Keiji Tsukada [1]

[1] Okayama University, 3-1-1 Tsushimanaka, Kita-ku, Okayama, Japan
[2] Superconducting Sensing Technology Research Association,
2-11-19 Minowa-tyo, Kouhoku-ku Yokohama, Japan
E-mail : pvpo2lvr@s.okayama-u.ac.jp

Abstract
Recently, the aging of buildings and infrastructure facilities is becoming a social problem. In the steel structure, the nondestructive evaluation with high lift-off is required to inspect the steel material under the asphalt or concrete, such as the steel floor slabs of roads. Under the high lift-off condition, the influence of the magnetic noise and applied magnetic field should be reduced because the signal from the crack is small. In order to reduce the influence of the magnetic noise, in this study, the detection of crack signal by eddy current testing was examined using a detection coil of a differential coil. This differential coil is a secondary differential coil combining two differential coils aiming for improving the signal to noise ratio (SNR) by reducing the influence of the magnetic noise and applied magnetic field. Using this secondary differential coil, the crack with high lift-off of 80 mm was detected, and it is expected to detect the cracks in steel structures located under the asphalt or concrete.

Keywords: NDE, crack, secondary differential coil, high lift-off

1. Introduction

As about 60 years have passed since the high economic growth period in Japan, the aging of metal parts such as the corrosion and cracks is becoming a serious problem [1], which are used in the highways, bridges and buildings. Thus, a technology to detect the aging of metal parts is required [2]. Some highways and bridges have steel plate under the asphalt or concrete, such as the steel floor slabs of roads. In the nondestructive evaluation (NDE) of highway and bridge, it is required to inspect the steel material with a distance between the sensor and steel structure (high lift-off). We have developed magnetic NDE technologies using the eddy current testing and the magnetic flux leakage testing [3-5]. In this study, the detection of crack with high lift-off was investigated by ECT using a copper coil as a detection coil. Under the high lift-off condition, the signal from crack is small, and the noise peculiar to the magnetic material and the applied magnetic field makes the signal detection difficult. Therefore, the influence of the magnetic noise and applied magnetic field should be reduced. In order to reduce the influence of the magnetic noise, a detection coil made of secondary differential coil combining two differential coils was developed and we attempted to detect a defect signal using this differential coil under the high lift-off condition. In addition, we investigated the optimal condition for crack detection by changing the relative angle between the coil unit and the crack.

2. Specific instructions

2.1 Secondary differential coil

In this study, the detection and induction coil were made of copper wire (Fig. 1). The inner diameter of induction coil is 100 × 160 mm with 50 turns (wire diameter: 0.4 mm), and the induction coil was also prepared by connecting two coils in reverse, and it can generate a dipole-like induced current under the center of the coil. The inner diameter of detection coil is

21 × 31 mm with 100 turns (wire diameter: 0.3 mm), and the secondary differential coil was prepared using two sets of differential coils as a detection coil. The detection coil is wound in the direction of the arrow shown in Fig. 1, and the coils connected in the order from 1 to 4 works as a secondary differential coil. By arranging two sets of differential coils in layers, it is possible to improve the signal to noise ratio (SNR) by reducing the magnetic noise and the influence of the applied magnetic field.

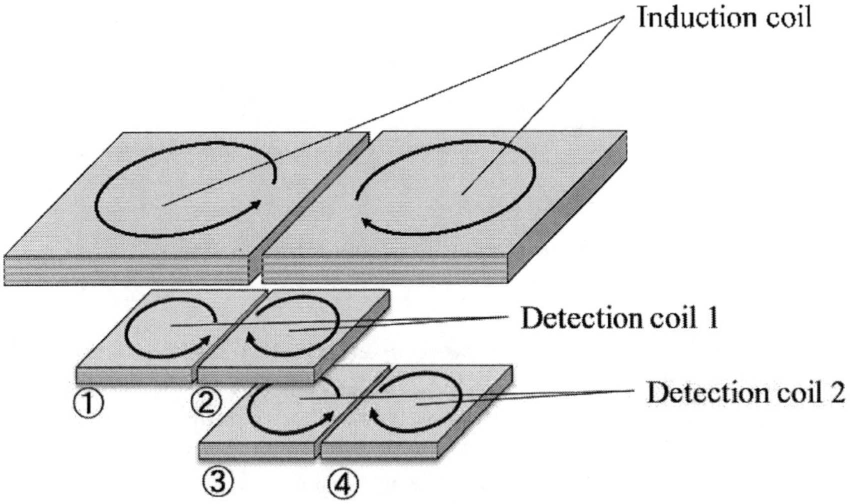

Fig. 1. Schematic of the coil unit.

2.2 Measurement system

Figure 2 shows the configuration of the measurement device used in this study. The current source was controlled by a function generator and the output of current source was applied to the induction coil. Then, the signal from the detection coil (secondary differential coil) was measured by a lock-in amplifier and the detected signal was saved on PC. At the same time, the position information where the magnetic field signal was obtained was also recorded by the logger. The induction / detection coil and various power supplies were implemented to a hand cart, and the guide rails were constructed on both sides of the sample, as shown in Fig. 3. This enabled to move the measurement system easily above the sample [6]. The lift-off can be changed by adjusting the height of the coil unit. The hand cart was moved from the edge to edge of sample (1 line scan), and the data of nine lines with 30 mm intervals was measured for imaging.

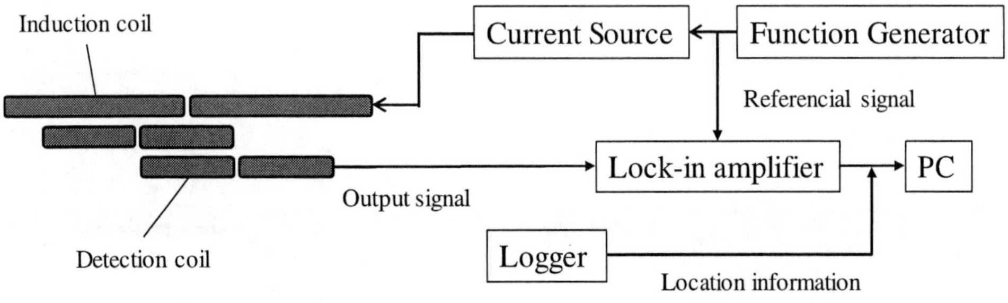

Fig. 2. Schematic of the measurement system.

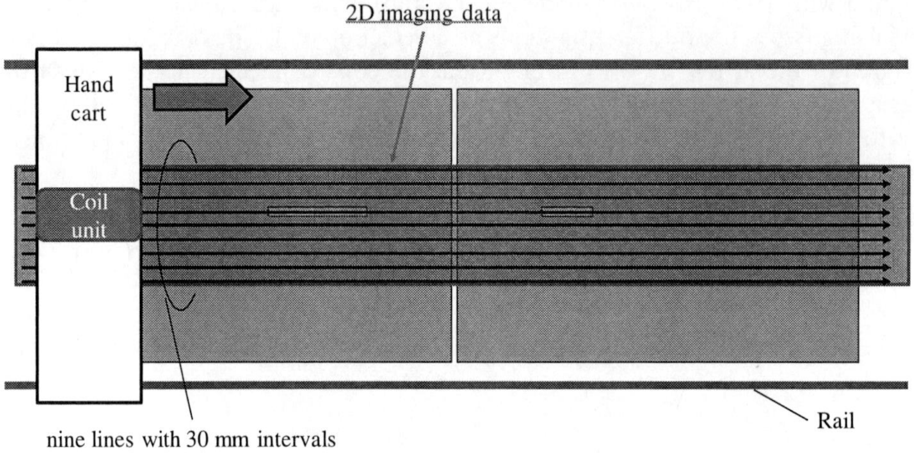

Fig. 3. Schematic of the system arrangement.

2.3 Measurement sample

Two steel plates (SM490A) with a crack were used as a test sample. The size is 0.7 m in width, 1.0 m in length and 6 mm in thickness. Each sample has surface cracks of 50 and 100 mm in size and 4 mm in depth. In addition, the distance between the sample and the detection coil was 80 mm. This lift-off (80 mm) is enough high considering the practical inspection because the typical thickness of asphalt pavement is approximately 50 mm. In addition, the angle dependence of crack and coil unit was evaluated by changing the angle of the induction coil to the slit. Furthermore, the relative angle between the induction coil and the detection coil was also investigated.

3. Results and Discussion

3.1 Coil angle to the crack

At first, the difference of the signal depending on the angle of the coil unit to the crack was evaluated. The induction and detection coils are arranged, as shown in Fig. 4 (Top view).
The signals of three angle patterns show in Fig. 4 were measured under the condition of 80 mm-lift-off, applied frequency of 1 kHz, and current amplitude of 1 A. Figure 5 shows each

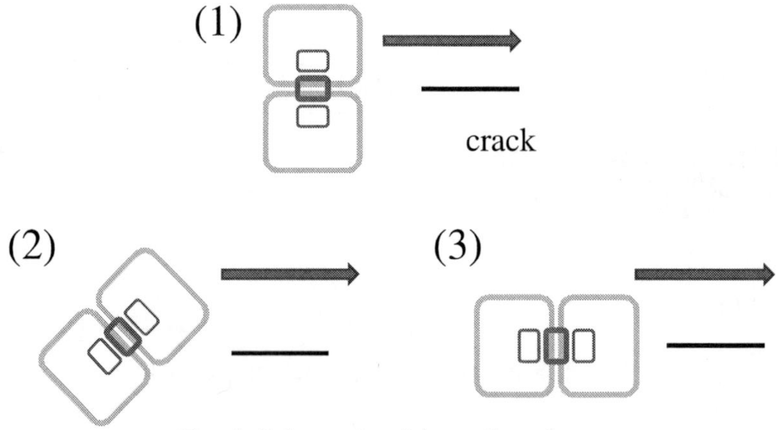

Fig. 4. Schematic of the coil angle.

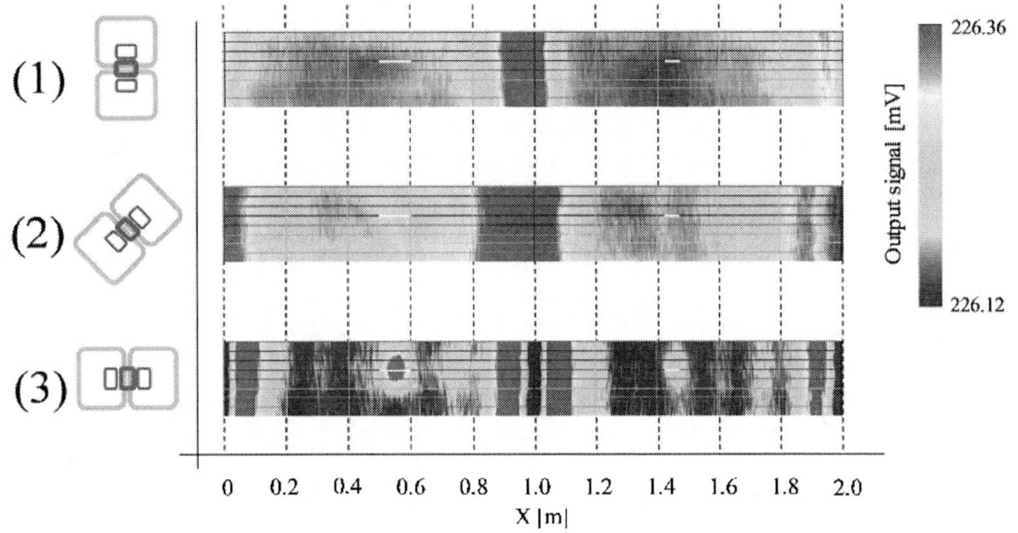

Fig. 5. Dependence of the coil angle.

mapping data. A sample with the crack of 100 mm was located on the left side and that of 50 mm was located on the right side.

From the result of Fig. 5(1), the signal change was not observed in the entire sample, and crack was not detected. When the signal was measured using the pattern (2), it seems that the signal changed slightly over a wide area around the crack compared to the case of pattern (1). However, it was difficult to detect the crack using the pattern (2). Finally, as can be seen in the pattern (3), clear signal changes above the cracks of both 50 and 100 mm were observed, and the crack could be detected unlike the patterns (1) and (2). From these results, it was found that the degree of crack detection depends on the angle of the coil unit. In the pattern (3), the signal changes above the crack, and the change is more clearly seen in the crack with a width of 100 mm. This is considered to be caused by the relative angle between the applied current dipole and the crack. To investigate the dependence of coil angle on the eddy current distribution, the eddy current distribution generated in the sample at angles (1) and (3) was analyzed by the finite element method. Figure 6 shows the eddy current distribution from just

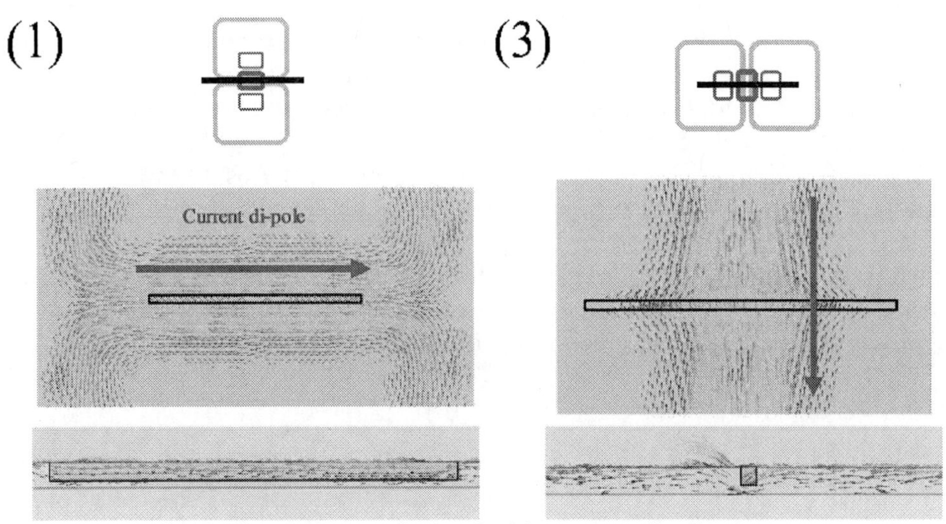

Fig. 6. Simulation analysis of eddy current distribution.

above the crack and that of the cross sectional view at the center of the crack.
In the pattern (1), the current dipole and the crack are parallel, while they are orthogonal in the pattern (3). It was found from the pattern (3) that the eddy current was generated so that the eddy current bypasses the crack and the eddy current flow changed near the crack. Because of this, the crack was detected. In the pattern (1), the current dipole and the crack are parallel to each other and the eddy current does not bypass like the pattern (3). As a result, the crack was not detected. From the above results and simulation analysis, the detection of the crack had an angle dependence of the coil unit, and the pattern (3) in which the applied current dipole and the crack are orthogonal to each other was most suitable for the detection.

3.2 Relative position of the coil

All experiments in Section 3.1 were conducted with the induction and detection coils were arranged as the pattern A shown in Fig. 7. To investigate the angle dependence of detection coil, the detection coil was rotated 90° from the arrangement of A, the measurement was performed in the arrangement of B, and the dependence of detection performance on the relative angle between the induction and detection coils was compared. The signal was measured under the three patterns of angles between the detection coil and the crack measured at 3.1, and a total of 6 patterns of measurement results were compared. The measurement conditions were the same as 3.1.

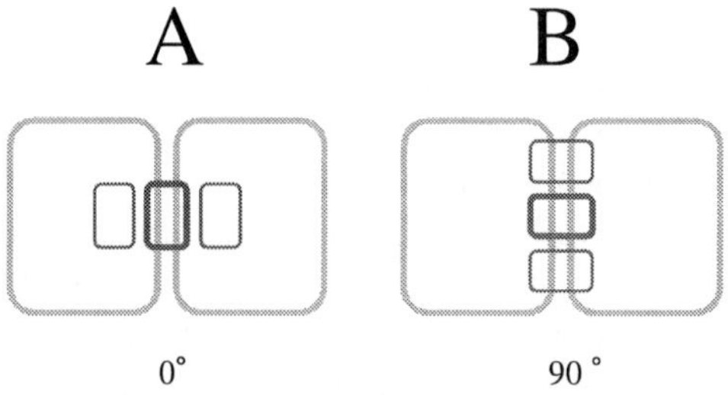

Fig. 7. Angle to the induction coil of the detection coil.

Figure 8 shows the measurement results for the three patterns of pattern B shown in Fig. 7. In Figs. 5(1) and (2), the crack could not be detected as described above, and in Fig. 8 (4), no signal change was also seen around the crack. Therefore, the detection of crack was difficult when the eddy current dipole is parallel to the crack. On the other hand, there were three patterns (Fig. 5(3), Figs. 8(5) and (6)) that could confirm the signal changes around the crack. In patterns of (3) and (5), the signal change was confirmed just above the crack. Although the crack of 50 mm was not detected in the pattern (6), a signal change was seen as the form of a quadrupole at the four corners of the crack. In these three patterns, the crack of both 100 and 50 mm were detected, and it was the pattern (3) that can detect the crack the most clearly. In addition, in the case of the pattern (5), the signal changes were seen at 100 and 50 mm although these changes were not as clear as the pattern (3). From this result, it was confirmed that when the coil unit with the arrangement of B was used, the detection was possible even when the current dipole and the crack were not orthogonal.

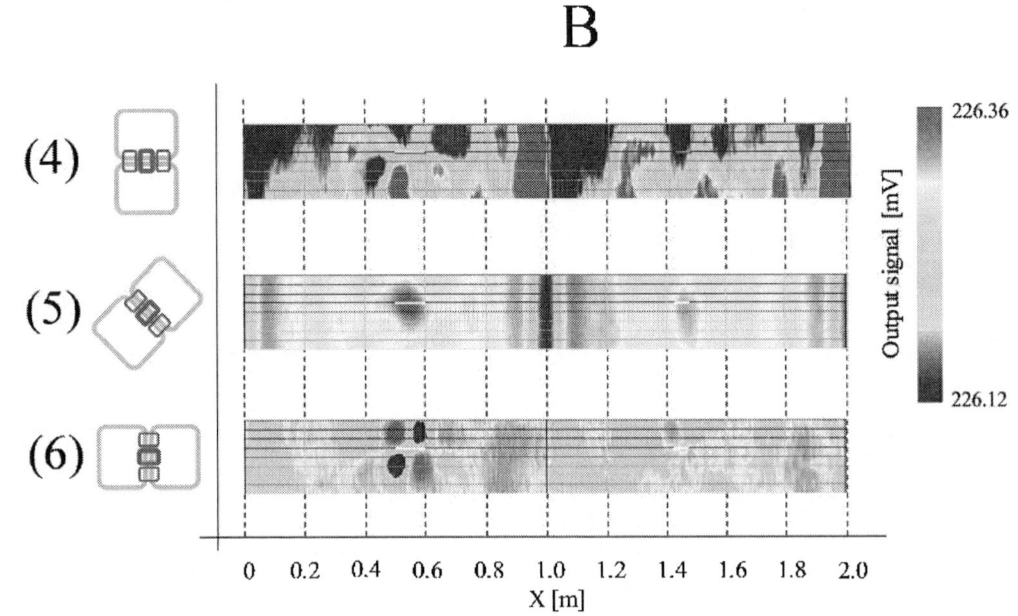

Fig. 8. 2D mapping image (3 pattern).

4. Conclusion

In this study, the magnetic noise and applied magnetic field were reduced by using a secondary differential coil as the detection coil, and slit-like cracks were detected with high lift-off. In addition, the degree of detection depended on the angle of the coil unit with respect to the crack. From the analysis of the measurement results and simulation, the arrangement in which the crack and the applied current dipole are orthogonal was the most sensitive, and detection was possible even at a high lift-off of 80 mm. By using this method, it is expected to detect the cracks in steel structures located under the asphalt or concrete.

References

1. T. Tamakoshi, M. Hoshino, A. Ichikawa, "Survey on Deterioration and Damage of Road Attachment Facilities", Technical Note of National Institute for Land and Infrastructure Management, no.685, 2012.
2. P. Kaviani, R. Anousheh, V. Carlos, "Fracture Critical Bridge Inspection", Structure magazine, pp. 12-14, 2016.
3. K. Tsukada, Y. Haga, K. Morita, N. Song, K. Sakai, T. Kiwa, W. Cheng, "Detection of inner corrosion of steel construction using magnetic resistance sensor and magnetic spectroscopy analysis", IEEE Trans. Magn., vol. 52, no. 7, 6201504, 2016.
4. K. Tsukada, T. Kiwa, T. Kawata, and Y. Ishihara: "Low-Frequency Eddy Current Imaging Using MR Sensor Detecting Tangential Magnetic Field Components for Nondestructive Evaluation", IEEE Trans. Magn., Vol. 42, 10, pp. 3315 – 3317, 2006.
5. K. Tsukada, T. Yasugi, Y. Majima, K. Sakai, and T. Kiwa: "Detection of Inner Cracks in Thick Steel Plates Using Unsaturated AC Magnetic Flux Leakage Testing With a Magnetic Resistance Gradiometer", IEEE Trans. Magn., Vol. 53, no. 11, 2501305, 2017.
6. A. Tsukamoto, T. Hato, S. Adachi, Y. Oshikubo, W. Cheng, K. Enpuku, K. Tsukada, K. Tanabe, "Eddycurrent Testing System Using HTS-SQUID With External Pickup Coil Made of HTS Wire", IEEE Trans. Appl. Supercond, vol. 27, no.4, 1600505, 2017.

Increased accuracy of spectral analysis in NDT

Karel HÁJEK

Department of Electrical Engineering, University of Defence, Brno, Czech Republic
Phone: +420 973 442 250, e-mail: karel.hajek@unob.cz

Abstract
Some NDT methods advantageously use spectral analysis using DFT. These are e.g. methods of nonlinear ultrasonic spectroscopy, resonance spectroscopy but also some others. Despite its indisputable advantages, the use of DFT spectral analysis brings some problems. A major problem is the leakage effect, which is evident in cases where we cannot provide a coherent DFT with synchronous sampling. Related to this is the problem of insufficiently accurate determination of frequencies of dominant harmonic components, which is manifested e.g. in detection of change of resonant frequencies during non-linear effects of tested systems. A common solution to these problems is the use of new methods for very accurate determination of the frequency of the dominant harmonic signal. This also allows the suppression of the leakage effect in spectral analysis, so that the spectrum obtained is accurate and without the leakage effect in a significantly higher dynamic range than the use of known window functions. The paper shows the possibilities of solving the mentioned problems.

Keywords: Spectral analysis, ultrasonic spectroscopy, resonance spectroscopy, DFT, windows

1. Introduction

DFT spectral analysis respectively its faster FFT variant is used in many technical fields. It has an important application in the area of some NDT methods. First of all we can mention some nonlinear and linear ultrasonic methods with spectral evaluation (e.g. [1], [2]) and acoustic resonance methods (e.g. [3], [4]). But it can be used for example for Doppler speed measurement (e.g. [5], [6]) etc. In addition to determining the resulting signal spectrum, it is also a common task to determine the frequency of the dominant harmonic component of the measured signal or to determine the transmission properties of the test system from discretized variables.

Of course DFT spectral analysis has its advantages and disadvantages. One of the main advantages is the standard acquisition of discrete data of the monitored variable obtained by a regular A-D conversion and the use of common computer technology by standard discrete signal processing (DSP) procedures. On the other hand, discretization of the monitored variables brings errors to the measurement and processing of the signal, such as the generally known leakage effect. Therefore, we strive to improve DSP methods to minimize these errors.

2. DFT of coherent and incoherent signals

Most of these methods of signal processing are based on the discrete Fourier transform (DFT) as presented in practically all basic DSP textbooks, e.g. [7], [8], [9], [10]. The basic definition is derived from the Fourier transform (FT) and is usually presented in the form

$$\dot{C}_n = \sum_{k=0}^{N-1} s(k) e^{-jn\frac{2\pi}{N}k} \quad . \tag{1}$$

However, it is advisable to start with the Fourier series (FS). Considering the known conditions of converting a continuous signal to a discrete one, such as satisfying the sampling

theorem and the periodicity of the discrete signal relative to the length N of the DFT window samples (coherent signal, $F_0=kF$, where k is an integer).
Then, the calculated FT values for the continuous signal and the values for the discrete Fourier series (DFS) can be matched as expressed by the following formulas for real, imaginary or complex spectrum components FT values for a continuous signal and values for a discrete Fourier series (DFS) as expressed by the following formulas for real, imaginary or complex spectrum components

$$A_n = \frac{2}{T}\int_T s(t)\cos(n\Omega t)\,dt = \frac{2}{N}\sum_{k=0}^{N-1} s(k)\cos\left(n\frac{2\pi}{N}k\right) \quad (2)$$

$$B_n = \frac{2}{T}\int_T s(t)\sin(n\Omega t)\,dt = \frac{2}{N}\sum_{k=0}^{N-1} s(k)\sin\left(n\frac{2\pi}{N}k\right) \quad (3)$$

$$\dot{C}_n = \frac{2}{T}\int_T s(t)e^{-jn\Omega t}\,dt = \frac{2}{N}\sum_{k=0}^{N-1} s(k)e^{-jn\frac{2\pi}{N}k} \quad (4)$$

Interestingly, these equalities are absolute. Obviously, if these conditions are met, the integral calculation of the coefficients A_n, B_n and C_n of the individual harmonic components for a continuous signal can be replaced by the sum of N terms for the sampled signal. This of course applies only to n in the interval 0 to $N/2$, which corresponds to the fulfillment of the sampling theorem.

Obviously, if the above conditions are met, three samples per period are sufficient to determine exactly the magnitude and phase of the sampled signal (identical to the FS of the continuous signal) regardless of the phase of the measured signal being sampled. This can be used e.g. in measuring systems, where we excite a known harmonic signal and sample with a frequency that is an exact integral N-multiple of the frequency of the measuring signal.

The time course and spectrum of such a signal is shown in Fig. 1. It is necessary to distinguish period T_0 and its corresponding frequency F_0 of the test signal, period T_S and its corresponding frequency F_S of the sampling signal. These values and the number N of samples in the DFT window indicate the period $T=NT_S$ and the corresponding frequency F of the first harmonic component of the resulting DFS spectrum. Obviously, the frequency of the test signal F_0 is an integral multiple of the frequency of the first harmonic component F (twice in Fig. 1). In this case, the resulting discrete spectrum of DFS consists of a single harmonic component with an amplitude K_n equal to the amplitude of the input continuous signal, where the other components have zero value due to zero transmissions of the corresponding FIR filters for given frequency values, see Fig. 1.

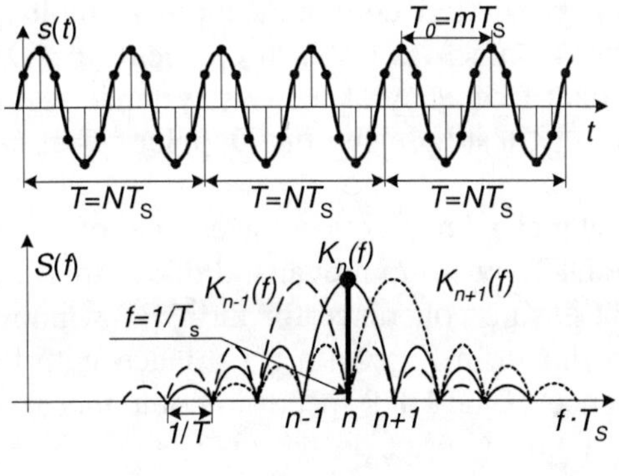

$$\frac{T}{T_0} = \frac{N}{m} = \frac{12}{6} = k = 2$$

Fig. 1 Time waveform and spectrum calculation of a discrete periodic signal (as a signal pass through FIR filters corresponding to DFS relations).

If the coherence condition ($T \neq kT_0$, $F_0 \neq kF$) is not met, the whole number of test signal periods will not fall within the DFS window period (see Fig. 2 a). In this case, the amplitude of the test signal express oneself not only for the two closest integral multiples of the first harmonic component (with FIR filter transfer <1), but also for other neighboring components with correspondingly decreasing FIR filter transmissions. Thus we obtain a much richer spectrum than the one-component spectrum of the tested signal (Fig. 2.c, Fig. 1). This phenomenon, which distorts the spectrum obtained, is known as "leakage" and one of the fundamental problems of DT spectral analysis.

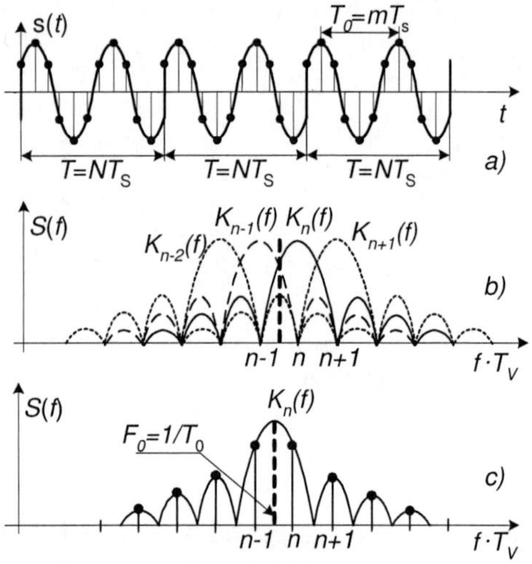

$$\frac{T}{T_0}=\frac{N}{m}=\frac{11}{6}=1{,}83\overline{3}$$

Fig. 2 Time course and spectrum of incoherent signal:

a) a) the time course of the equivalent signal for the selected period T,

b) b) calculation of one component of the spectrum of the incoherent signal as a signal pass through FIR filters,

c) c)an equivalent expression of the calculated spectrum using one magnitude response shifted to the real signal frequency (leakage effect).

It is also necessary to explain the difference between the definition of DFT (1) and DFS (4). Obviously, this difference formally resides in the *2/N* term, which is applied to DFS in spectrum calculation, while DFT is used in backward Fourier transform. This seemingly formal difference has some practical disadvantages in terms of FTT. Importantly, the calculated magnitude of the spectral components does not match the amplitudes of the corresponding continuous harmonic signals and varies depending on the N value. Therefore, in practice it is often used de facto DFS, although it is referred to as DFT resp. FFT.

3. Frequency estimation of dominant harmonic signal

One of the practical tasks of signal spectrum analysis is to estimate the frequency of the dominant harmonic signal if its exact value is unknown and is needed to determine the important properties of the tested signal or the tested system. In technical practice, it is used in a very wide range of tasks, for astronomy, sonar, radar and communication technology, acoustic technology and music, speech analysis after as much as GPS. In the field of NDT, this task is solved eg for accurate detection of resonant frequency of various systems, e.g. in acoustic resonance method or as impact echo. Similarly, testing using the Doppler effect can be used.

This task can be solved in both time and frequency planes. In practice, the estimation of the frequency of the dominant harmonic signal with the DFT spectrum prevails. Since in this case we do not know the detected frequency, we cannot ensure coherence for the DFT window. The initial conditions for solving this problem are shown in Fig. 3, in accordance with the previous chapter, where an example of the dependence of the DFT spectrum components on the relative frequency $k=F_0/F$, for which bin designation is used. In the case of a coherent signal, the frequency would correspond to an integer value of k. In our case of an incoherent signal, the relative value of the actual frequency k_0 is in the range of k to $k+1$ which correspond to the two highest components of the X_k and X_{k+1} DFT spectrum.

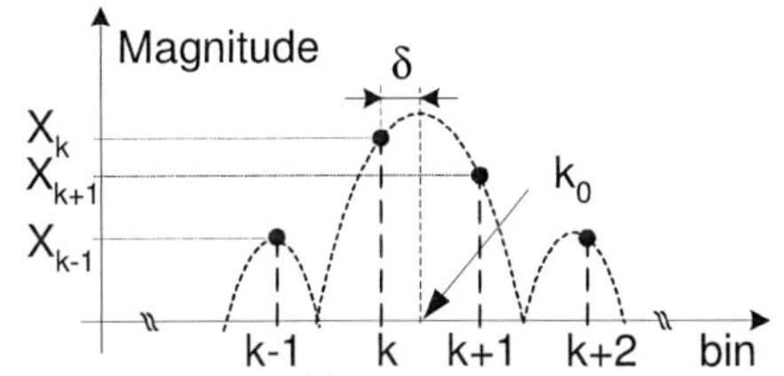

Fig. 3 Highest amplitude DFT spectrum components used to calculate an estimate of the signal frequency using the deviation value δ.

The relative frequency of the actual component k_0 is shifted from the relative frequency k of the highest component by a shift δ, the value of which can be in the range -0.5 to 0.5. Therefore, as a basic task of estimating the frequency of a dominant signal, it is converted to a value determination. A number of methods based on different approaches and with different complexity have been published to solve this task. Depending on the method approach and parameters such as N and the F0 / FS ratio, different estimation accuracy is achieved. Methods can be divided into two basic groups, simple and complex.

Simple methods usually involve the direct application of a relationship that determines the value δ in one or two steps of the procedure. More complicated methods achieve higher estimation accuracy than the previous group of methods, but usually have to use a much more sophisticated iterative algorithm. In common practice, simple methods are more widely used, as they usually achieve sufficient accuracy with low computational complexity.

The most commonly used simple methods are usually based on the use of the values of the three highest components of the spectrum (X_{k-1}, X_k, X_{k+1}), see Fig. 3. Of these, the value can be expressed simply by a known and often stated relationship

$$\delta = \mathrm{Re}\left[\frac{X_{k-1} - X_{k+1}}{2X_k - X_{k-1} - X_{k+1}}\right], \tag{5}$$

see [11], resp. slightly refined by [12]. The accuracy of this estimate is not very high and corresponds approximately to the accuracy that is similarly limited by the use of an 8-bit AD converter.

To increase the accuracy of the estimation, an auxiliary step is used. Before calculating the DFT spectrum we apply some of the known window functions to suppress the aliasing, which arises by applying a rectangular DFT window, to N samples of the input signal. The Hann window (symmetrical variant) has proved to be very advantageous. For the purpose of this article, the relationship for the DFT spectrum adjusted by this Hann window was derived

$$\delta = \mathrm{Re}\left[2\frac{X_{k-1} - X_{k+1}}{2X_k - X_{k-1} - X_{k+1}}\right], \tag{6}$$

where coefficient 2 corrects the effect of using Hann's window in comparison with (5). Thanks to the use of Hann's window, this estimate achieves a significantly higher accuracy than formula (5), as shown in the comparison in Fig. 4.

Fig. 4 shows the dependence of the root mean square error (RMSE) of value δ on both F_0/F_S and N. The symmetry is evident in terms of F_0/F_S with a minimum for F_0/F_S=0.25. Also the error monotonously falls with increasing of N. However, it can be stated that exist also other procedures and relationships based on the use of Hann's window, e.g. [13], which reaches approximately the analogous accuracy as that relationship (6). The precision of procedures with Hann's window, apart from the boundary values of the F_0/F_S ratio, achieves a relatively high accuracy, corresponding to approximately a 16-bit converter.

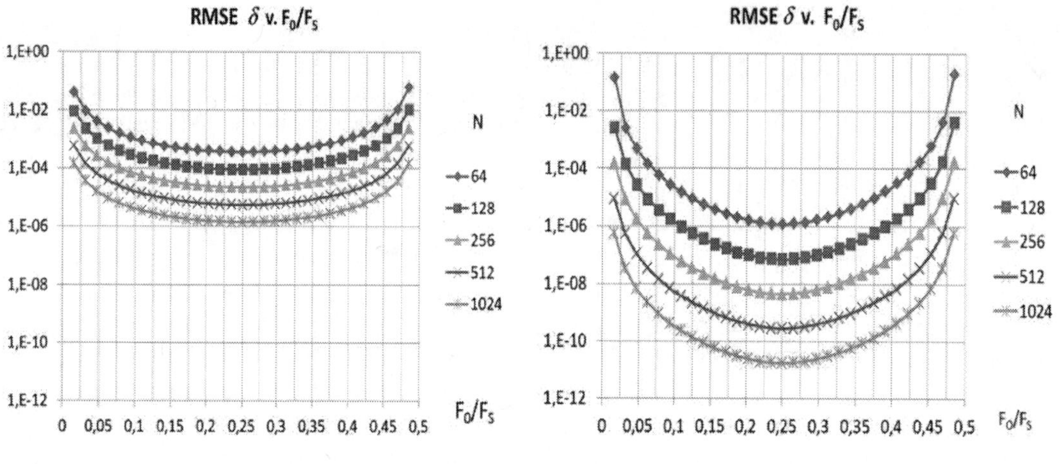

Fig. 4 RMSE of δ vs. $64\, f_0 / f_s$ for $\delta = 0.25$ for various N: a) estimator (5), b) estimator (6).

The dependence of the error on the value N in comparison for both cases and for the selected values δ and F_0/F_S for the ideal noise-free signal is shown in Fig. 7a). It is also interesting to discuss the influence of signal noise on the accuracy of frequency estimation. The dependence of the error on the signal-noise-ratio (SNR) value for both cases and for the chosen values $\delta = 0.25$, $F_0/F_S = 0.1875$ is shown in Fig. 7a). It is apparent that for the signal obtained by the 8-bit A-D conversion, the use of the first method is sufficient. However, for signals obtained by multi-bit A-D conversion (e.g. 12 or 16 bit), it is necessary to use the second estimation method according to (6). In doing so, the absolute accuracy of the estimation is limited by the theoretical Cramer-Rao bound (CRB) [14].

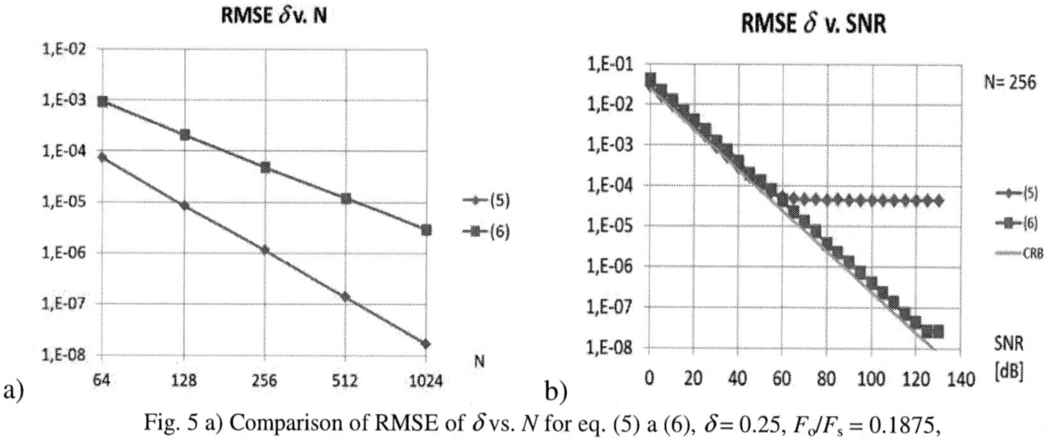

Fig. 5 a) Comparison of RMSE of δ vs. N for eq. (5) a (6), $\delta = 0.25$, $F_o/F_s = 0.1875$,
b) comparison of RMSE vs. SNR for eq. (5) a (6), $\delta = 0.25$, $F_o/F_s = 0.1875$, $N = 256$

4. DFT spectral analysis with suppressed leakage effect

As already mentioned, classical DFT spectral analysis has the main practical limitation caused by the leakage effect. In the current practice, this effect is reduced by using window functions. Conventional DFT spectrum analyzers usually offer a few special windows in addition to the basic rectangular window, eg Hamming, Hann, Triangle, Bartlet etc. These have different degrees of suppression of the leakage effect for more distant components. However, this is paid by expanding the band of the closest components in the so-called first lobe around the dominant component. This is a well known theory and practice in the use of DFT spectrum analyzers. However, using one of these windows only reduces the leakage rate and for many applications it does not address it sufficiently. Therefore, there is a new solution that allows practically complete suppression of this negative effect.

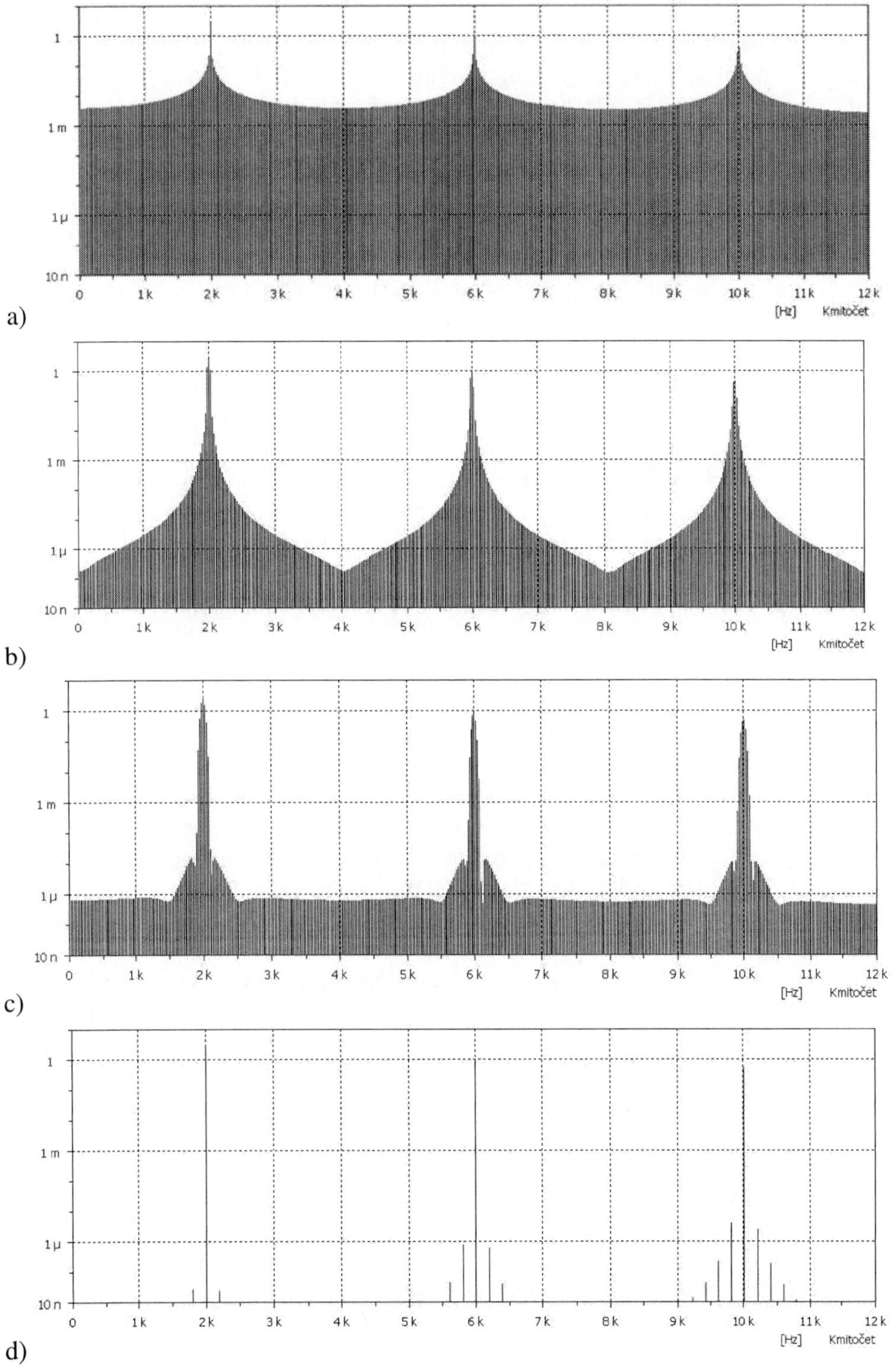

Fig. 6 Examples of rectangular DFT spectra with varying degrees of leakage suppression: a) rectangular window, b) Han window, c) Blackman-Harris window, d) use of the new method.

The basic idea is that a coherent signal does not exhibit a leakage effect. Therefore, this effect can be suppressed by changing the incoherent signal to coherent signal. We achieve this by calculating the inaccurate DFT spectrum with the leakage effect after the initial sampling of the N points of the test signal and calculating the deviation value d from it. Based on this value, we can calculate what the sampling frequency should be to create a coherent signal. This will allow us to obtain a coherent form of the original signal by some of the known resampling methods. By a second DFT calculation we can obtain a clear spectrum of the signal without the leakage effect. This is shown by the DFT spectrum of a rectangular signal modeled without the effect of A-D conversion for a more general comparison of the individual methods of the suppressing of leakage. Fig. 6a shows the spectrum obtained using a conventional rectangular window. The leakage effect here is very pronounced and limiting the accurate evaluation of the actual spectrum. Fig. 6b) shows the spectrum of the same signal with using the Hann window. This has to some extent reduced the leakage effect mainly for the more distant components from the dominant components. Fig. 6 (c) shows the use of a Blackman-Harris window, which effectively suppresses leakage over almost the entire band excluding the nearest components. On the contrary, they increase, which is obvious, for example, compared to a rectangular window.

The spectrum obtained using the new method shows almost absolutely accurate results. Small parasitic components are due to oversampling process errors. Thus, in the second approach, if the next series of N points were actually sampled at a precisely set sampling rate, interpolation errors would not occur there. On the other hand, it should be considered that using a real A-D converter would limit the purity of the spectrum by corresponding noise (see Figure 5b).

5. Conclusion

The paper shows the use of new knowledge in the field of DFT spectral analysis for some NDT methods. First, it explains the advantages of using DFS over DFT. It shows the suitability of using synchronous sampling for measuring some quantities in NDT. It also shows the known problem and causes of leakage for incoherent signals.

On the basis of this analysis, it shows ways to solve the problem of the most accurate estimation of the frequency of the dominant harmonic signal. The proposed method of (6) gives very accurate estimation values that are sufficient for most practical tasks and for the NDT field.

Furthermore, it shows a relatively simple possibility of suppressing the leakage effect in DFT spectral analysis, which is considerably more efficient than using standard window functions and makes it possible to obtain a spectrum practically without leakage.

Acknowledgements

For research, the infrastructure of Department of Electronic, University of Defence Brno, was used.

References

1. J. Kober, Z. Prevorovsky, 'Theoretical investigation of nonlinear ultrasonic wave modulation spectroscopy at crack interface', NDT & E International Volume 61, January 2014, Pages 10-15
2. M. Meo, U. Polimeno, G. Zumpano, 'Detecting Damage in Composite Material Using Nonlinear Elastic Wave Spectroscopy Methods', Appl Compos Mater (2008) 15: 115. https://doi.org/10.1007/s10443-008-9061-7
3. Polimeno, U., Meo, M., Almond, D.P. et al. 'Detecting Low Velocity Impact Damage in Composite Plate Using Nonlinear Acoustic/Ultrasound Methods', Appl Compos Mater (2010) 17: 481. https://doi.org/10.1007/s10443-010-9168-5
4. A. Novak, M. Bentahar, V. Tournat, R. El Guerjouma L. Simon, 'Nonlinear acoustic characterization of micro-damaged materials through higher harmonic resonance analysis', NDT & E International, Volume 45, Issue 1, January 2012, Pages 1-8
5. Thanh T. Nguyen ; J. C. Devlin ; D. M. Elton ; G. Deng ; E. Custovic ; B. Bienvenue, 'A comparison of FFT-based techniques for Doppler velocity estimation in SuperDARN radars', 2014 International Symposium on Information Theory and its Applications, December 2014, ISBN: 978-4-8855-2292-5
6. G. Sattibabu, Ch. Sridevi, S. Phani, G. Sridevi, 'Design of Velocity-Measuring System with Doppler Radar Concept and FFT Algorithm Based On ARM Processor for Traffic Safety', International Journal of Advanced Research in Electronics and Communication Engineering (IJARECE) Volume 1, Issue 4, October 2012
7. A. V. Oppenheim, and R. V. Schafer, 'DFT Analysis of Sinusoidal Signals' in Discrete-Time Signal Processing (3rd Edition), Prentice-Hall, 2010, ISBN: 978-0131988422
8. R. G. Lyons, 'The DFT Frequency Response to a Real Cosine Input' in Understanding Digital Signal Processing (2nd Edition), Prentice Hall, 2004, ISBN: 0131089897
9. S. W. Smith, 'Applications of the DFT' in The Scientist and Engineer's Guide to Digital Signal Processing. California Technical Publishing, 2003, ISBN: 978-0966017632
10. M. J. Roberts, 'Sampling and Discrete Fourier Transformation in Signals and Systems: Analysis Using Transform Methods and MATLAB®', McGraw Hill, 2004, ISBN: 0072499427
11. E. Jacobsen and P. Kootsookos, 'Fast, accurate frequency estimators,' IEEE Signal Process. Mag., vol. 24, pp. 123–125, May 2007.
12. Candan, C. (2011). 'A Method For Fine Resolution Frequency Estimation From Three DFT Samples'. IEEE Signal Processing Letters, 18(6), 351–354
13. D. Agrež, 'Weighted Multipoint Interpolated DFT to Improve Amplitude Estimation of Multifrequency Signal', IEEE Trans. Instrum. Meas., vol. 51, no. 11, pp. 287–292, April 2002, DOI: 10.1109/19.997826.
14. S. M. Kay, 'Fundamentals of Statistical Signal Processing, Volume 1:Estimation Theory.' Upper Saddle River, NJ, USA: Prentice-Hall, 1993.

Acoustic emission sources from fast dislocation motion

Petr HORA [1], Anna MACHOVÁ [1], Jan ČERV [1], Alena UHNÁKOVÁ [1]

[1] Institute of Thermomechanics of the Czech Academy of Sciences v.v.i.; Prague, Czech Republic

e-mail: hora@cdm.cas.cz, machova@it.cas.cz, cerv@it.cas.cz, uhnakova@it.cas.cz

Abstract

Acoustic emission from the fast dislocations emitted from an edge crack in 3D bcc iron crystal is studied via atomistic simulations by molecular dynamics technique. Acoustic emission patterns arising from the fast dislocation motion in molecular dynamics are visualized via the local kinetic energies of individual atoms and further modeled as a moving source of the stress waves in the anisotropic continuum.

Keywords: acoustic emission, edge crack, dislocation emission, free surface, bcc iron

1. Introduction

We studied motion of the dislocations <111>{112} emitted from the crack tip in a 3D bcc iron crystals. We maped the atomic positions of the dislocations by means of coordinates in the atomic lattice and as well via maping of the local kinetic energies of the individual atoms. We have found that in bulk of the crystal, the dislocation motion is subsonic (i.e. $V_{disl} < C_T$ <111> = 3007 m/s), while near the free surface (001) the dislocations accelerate to transonic regime $V_{disl} > C_T$ <111>.

There are two possible sources for the acceleration: i/ so called mirror attractive surface forces proportional $1/R$ where R is the distance of the dislocation core from the free surface; ii/ initial surface stress caused by surface relaxation of the crystal. The surface relaxation is observed as well experimentally. The paper [1] shows that the transonic regime can be caused by the short ranged surface stress arising after surface relaxation. This contribution is focused to acoustic emission caused by the fast dislocation motion in the atomistic lattice of a 3D bcc iron crystal and as well to modelling of this phenomena in an adequate anisotropic continuum.

2. MD simulations

A pre-existing through edge crack of the length a is considered, placed in the middle of a rectangular sample of length L, width W and thickness B - see Fig. 1. The relative crack length corresponds to $a/W = 0.8$. The sample was loaded by monotonic uni-axial tension (mode I) via prescribing displacements in L- direction for 6 border BW layers at the bottom and the top of the sample. Surface relaxation was performed prior to the loading.

Newtonian equations of motion for individual atoms were solved in all 3 directions by a central difference method, using time integration step $h = 1\text{x}10^{-14}$ s. The surface region mentioned, with prescribed displacements, does not enter the time integration procedure. No periodic boundary conditions were applied in the free 3D atomistic simulations presented.

Atomically sharp edge crack $(\bar{1}10)[110]$ was created by cutting interatomic bonds across the initial crack plane. The initial temperature was 0 K and further thermal atomic motion was not controlled, similar to [1-2]. The atomistic sample consisted of 440 atomic planes $(\bar{1}10)$ in L-direction, 220 planes (001) in W-direction and 30 planes (110) in B-direction along the crack front.

The total number of atoms in the 3D crystal was $NPOIN$ = 1 452 000. The atomistic specimen of SEN (single edge notch) type was loaded gradually (linearly in time $t = n\,h$) with prescribed displacements $v(t)$ at the BW surfaces each time step $t = n\,h$ in the direction $[\bar{1}10]$ along the sample length L, i.e. in mode I.

The coordinates, the local kinetic energies $E_{kin}(li)$ and the local number of interactions KNT (i.e. coordination numbers) at individual atoms (li) were monitored at selected time steps for graphic purposes in 3D.

For MD simulations we use our in-house fortran codes and, for the graphical treatment of MD results, our codes use the commercial computing environment MATLAB.

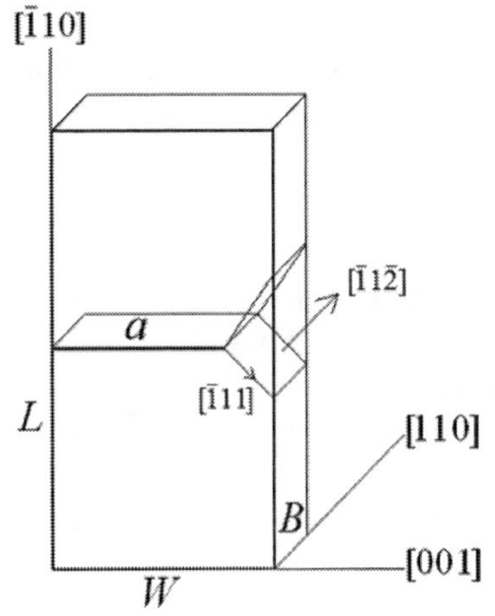

Figure 1. Sample geometry for crack orientation $(\bar{1}10)[110]$ (crack plane / crack front) and scheme for the available inclined slip systems: the arrow $[\bar{1}11]$ denotes the slip direction, the arrow $[\bar{1}1\bar{2}]$ marks the normal to the lower slip plane $(\bar{1}1\bar{2})$ containing also the crack front oriented in the [110] direction. The upper slip system is of orientation $[1\bar{1}1](\bar{1}12)$.

3. Results and discussion

Here we present the results for a relatively slow loading rate $\Delta v_A(t)/\Delta t$ = 14.7 Å / 20 000 h, where v_A is the applied prescribed displacement. The critical prescribed displacement v_G = 7.35 Å (corresponding to the critical Griffith stress intensity K_G) is reached per 10 000 h. The loading rate may also be expressed via K_G as dK_A / dt = 0.010 K_G / ps (since 100 h = 1 ps). Here, K_G = 0.906 MPa m$^{1/2}$ stands for Griffith stress intensity [3-4] needed theoretically for cleavage crack extension of the crack $(\bar{1}10)[110]$ with the interatomic potential in use.

In our case the available slip systems (Fig. 1) are <111>{112}, their inclination angle α with respect to the axis of potential crack extension is about 55°, the so-called Schmid factor is $\sin\alpha\cos\alpha = 0.47$ and K_{disl} = 0.892 MPa m$^{1/2}$ according to Beltz-Rice model [3-4] including beside the K-factor also the T-stress.

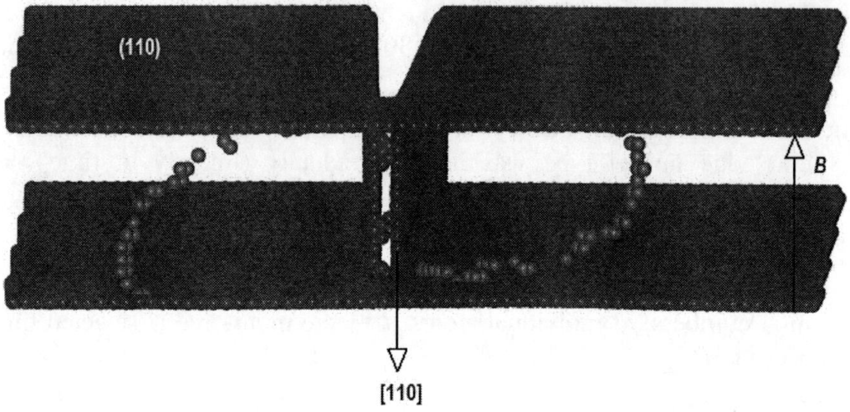

Figure 2. 3D visualization of the first emitted dislocations at the edge crack $a/W = 0.8$, time step 11 861, $v_A(t) = 8.72$ Å. Dislocation cores are shown by means of the red atoms (KNT = 16), the crack surfaces and the free sample surfaces are shown via blue atoms (KNT = 10). The arrow [110] shows orientation of the crack front and of the crystal thickness B (see Fig.1).

At time step 10 760 ($K_A(t) = 1.076\ K_G$), we detected the first increase in number of interactions *LINT* in MD, indicating the beginning of the dislocation generation, followed by subsequent emission, as presented in the next figures. The complete pair dislocation emission occurred [1] at time step 11 600. Fig. 2 at later time 11 861 illustrates that curved dislocations (red atoms) of mixed character with screw and edge components are emitted in the 3D atomistic sample of finite dimensions. The atoms in Fig. 2 were selected according to the coordination numbers *KNT*. The regular coordination number in the perfect bcc lattice corresponds to $KNT = 14$. (Note that *KNT* may differ for different interatomic potentials.)

In the middle of the crystal (at $B/2$ in Fig. 1), plane strain conditions prevail at the crack front and so, the dislocation are of edge character. At the free *LW* surfaces, where plane stress conditions must be obeyed in z – direction, the screw character of dislocation prevails.

Some atomistic studies utilize periodic boundary conditions along the crack front to model a thick specimen in z-direction. In such a case an emission of straight (line) dislocations from the crack front is expected, similar to plane strain (2D) simulations as in [4]. Here we do not use the periodic conditions in 3D crack simulations since they do not enable generally free dislocation emission, e.g. from oblique slip systems.

The emission of the complete edge dislocations <111>{112} is shown in Fig. 3a. The emission is complete when the relative shear displacement U10 at the crack tip (see [1]) reaches the value $b = a_0\sqrt{3}/2$, where ao means the lattice parameter and b is the Burgers vector in the bcc lattice. It leads to crack blunting and to a significant decrease of the potential energy in the system. As shown in [1], during the emission, a quick transition from the configuration $b/2$ to b decreases the high shear stress induced by the crack which leads to radiation of the stress waves (acoustic emission (AE)) shown in Fig. 3b via the map of the local kinetic energies $E_{kin}(li)$ of individual atoms li.

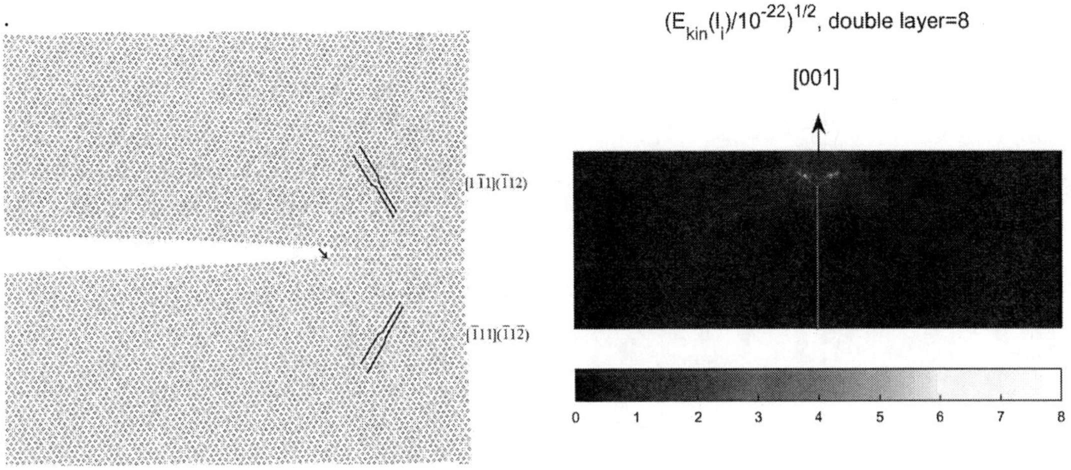

a) b)

Figure 3. Visualization of the first dislocation emission on the inclined slip systems <111>{112} at the edge crack $a/W = 0.8$ in the middle of the crystal, time step 11 600 ($K_A(t) = 1.16\ K_G$) $v_A(t) = 8.53$: a) by means of atomic coordinates; b) by means of local kinetic energies $E_{kin}(li)$ of individual atoms.

The value 8 on the colour scale corresponds to a maximum of the local kinetic energy in the crystal; at the bright dislocation cores $E_{kin}(li) = 0.33 \times 10^{-20}$ J, while the average value per one atom in the crystal is $EKIN\ /\ NPOIN = 0.10 \times 10^{-22}$J. Further, at the bright dislocation cores $E_{kin}(li) = 0.33 \times 10^{-20}$ J, while the average value per atom in the atomistic sample is $EKIN\ /\ NPOIN = 0.10 \times 10^{-22}$ J, which on the color scale corresponds to 0.01, i.e. to the black color. In the case of dislocation cores it is $\sqrt{33} = 5.7$ and their color is yellow. In terms of the Maxwell-Boltzmann relations for 3 degrees of freedom, the kinetic energy at the dislocation core corresponds to the temperature of about 159 K, while the average temperature in the atomistic sample (Fig. 1) is 4.8 K.

Motion of the individual dislocation segments may differ in 3D. Here we will follow and evaluate the generation and kinetics of motion of the lower edge dislocation lying below crack faces in atomistic Figs. 3a, 5a, which is the right dislocation in Figs. 3b, 5b, by the map of the local atomic kinetic energies. In this case, Burgers vector lies in the observation plane (110) and the direction [110] of the dislocation line is parallel with crack front and perpendicular to the observation plane – see the precise notation in Fig. 3a) and Fig. 5b).

Atomic monitoring of the dislocation motion in a middle plane (110) (at $B/2$) is shown in Fig. 4 by means of the dimensionless positions $r\ /\ b$ of the lower dislocation in dependence on dimensionless time $\Delta t/h$, where $\Delta t = t - 11\ 600\ h$. The line corresponds to an average subsonic dislocation velocity of $V_{disl} = 2210$ m/s ± 407 m/s that was obtained by the least-square method in the time interval <11 600 h, 11 880 h>, or according to Fig. 4 in the interval $\Delta t/h \in$ <0, 280>. In the time interval <11 880 h, 11 900 h>, i.e. close to the free surface, the slope of the curve differs, as may be seen in Fig. 4 for $\Delta t/h \in$ <280, 300>. The dislocation in MD accelerates at a very close distance of about $4b$ from the right free surface. It leads to an instantaneous dislocation velocity in the <111> direction, either transonic $V_{disl} = (46\text{-}43)\ b\ /\ 20h = 3724$ m/s above $C_T = 3007$ m/s [1], or to supersonic $V_{disl} = (46\text{-}43)b/10h = 7447$ m/s above $C_L = 6487$ m/s [1], if we suppose that the surface step was already created at the time step 11 890.

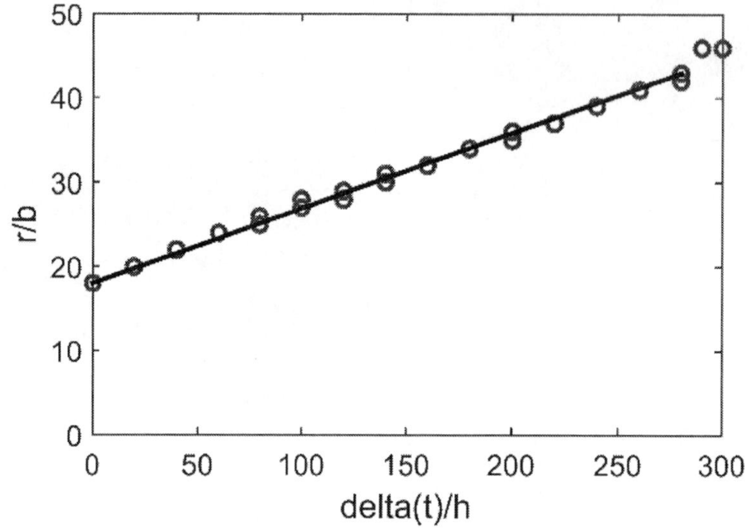

Figure 4. Dislocation motion by means of dimensionless position r/b from crack faces in dependence on dimensionless time $\Delta t/h$, where $\Delta t = t$ -11 600 h. The line corresponds to an average dislocation velocity $V_{\mathrm{disl}} = 2210$ m/s ±407 m/s.

Arrival of the transonic dislocations to the free right *BL* surface (001) is shown in Fig. 5 for time step 11 885. The dislocation cores are visible as the light “hot” points at the free surface.

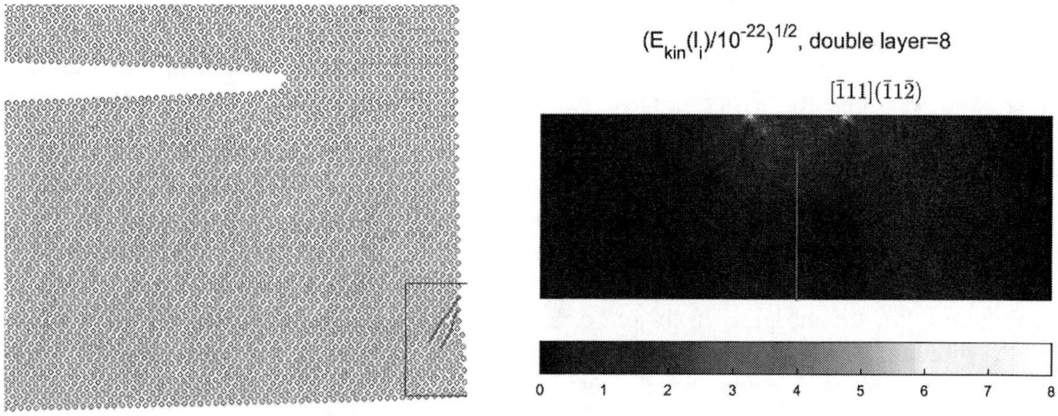

a) b)

Figure 5. Arrival of the emitted dislocations to the free sample surface, middle of the crystal, time step 11 885: a) the complex (lighted) atomic configuration before arrival of the dislocation to the free surface; b) acoustic emission signal from the transonic dislocation motion by means of the local kinetic energies.

The dislocations in Fig. 3a) at time step 11 600 begin to move away from the crack tip and the right free surface does not still influence significantly the AE signal in Fig. 3b). The character of the AE signal in Fig. 3b) is predominantly given by the envelope of the shear waves emitted from the moving dislocation cores in the two active slip systems <111>{112}, similar to [5].

The character of AE signal from the dislocations approaching the free surface in Fig. 5b at time step 11 885 is different. In this case, the AE patterns are already strongly influenced by the near free surface. Dislocation cores in Fig. 5b) are visible by means of the yellow "hot" points. When the dislocations penetrate the right free surface layers, one may observe that the atomic configuration in Fig. 5a) does not well define the position of the dislocation core, as the core is being split and looks like a collective motion of atoms, which creates a surface step. In comparison with Fig.3b), the envelope of the shear waves emitted in Fig. 5b) is more separated, via the dark region (with a low kinetic energy), from the entire wave region generated during the previous dislocation motion. This indicates an acceleration of the dislocations close to the free surface.

In Fig. 6 we present simplified reconstruction of the acoustic emission from the transonic dislocation motion near by the free surface in the framework of anisotropic continuum of our crystal symmetry. For the potential used, the velocity of the transversal (shear) waves in the <111> direction corresponds to C_T = 3007 m/s and for the longitudinal waves it is C_L = 6487 m/s [1], as mentioned above. In anisotropic medium, only some crystallographic directions (in our case <100>, <011> and <111>) transfer pure longitudinal or pure transversal mode, in other directions the elastic waves propagate as a quasi-longitudinal (qL) or quasi-transversal (qT) waves [5]. Note that in [5] we treated dislocation emission from a short central crack in an extended crystal and the influence of free surfaces was not examined in [5]. Fig.6a shows the section of the ray (wave) surfaces using the group velocities of the quasi-longitudinal qL and quasi-transversal qT (shear) waves in our observation plane (110) in the bcc iron crystal under plane strain conditions. The point source of the anisotropic AE signal lies in the middle of Fig. 6a and may represent e.g. the crack tip point or dislocation core. The axis of the potential crack extension in Fig. 6 and Figs. 3b, 5b corresponds to [001] direction, see Fig. 1. As mentioned above, our slip system <111>{112} for dislocation emission is oriented under an angle of about 55° ($\tan(\alpha) = \sqrt{2}$) with respect to the [001] direction of the potential crack growth. The dislocation emission from the crack tip is a shear process and so it can be modeled as a moving point source of the shear qT waves along the slip direction <111> inclined under ~55° with respect to [001] direction – see Fig. 6b), where the moving point source is denoted via (+) for a dislocation velocity of about 1.36 C_T. The crosses in the <111> direction correspond to Burgers vector. The envelope of the shear qT waves emitted in Fig. 6b) represents an AE signal arising from the transonic motion of the right (lower) emitted dislocation in an extended (unbounded) observation plane (110), i.e. it shows the situation during dislocation motion in the interval from the position 42*b* up to 46*b*, unimpeded by the free surface and by the back wave reflections from free surface. If we perform in Fig. 6b) a cut selection near by the last position (46 *b*) of the dislocation core (•) to model the arrival of the dislocation close to the free surface, we obtain in Fig. 6c an AE signal similar to Fig. 6d coming from the transonic motion of the right dislocation approaching the free (001) surface in MD at time step 11 885 (Note that Fig. 6d presents a detail from MD results in Fig. 5b, performed with a gray color scale in MATLAB).

The simplified reconstruction in Fig.6 does not include the back wave reflections acting in MD and the previous radiation coming from dislocation generation and subsequent subsonic dislocation motion. Nevertheless, the shape of AE signal and Mach angles of about 65° following from the wave reconstruction in Fig. 6c roughly describes the independent MD signal in Fig. 6d at the free (001) surface. It implies that the strongest AE signal from the crack induced dislocation emission is transferred via qT waves, which is a useful information for the experimental AE detection with the velocity sensitive transducers.

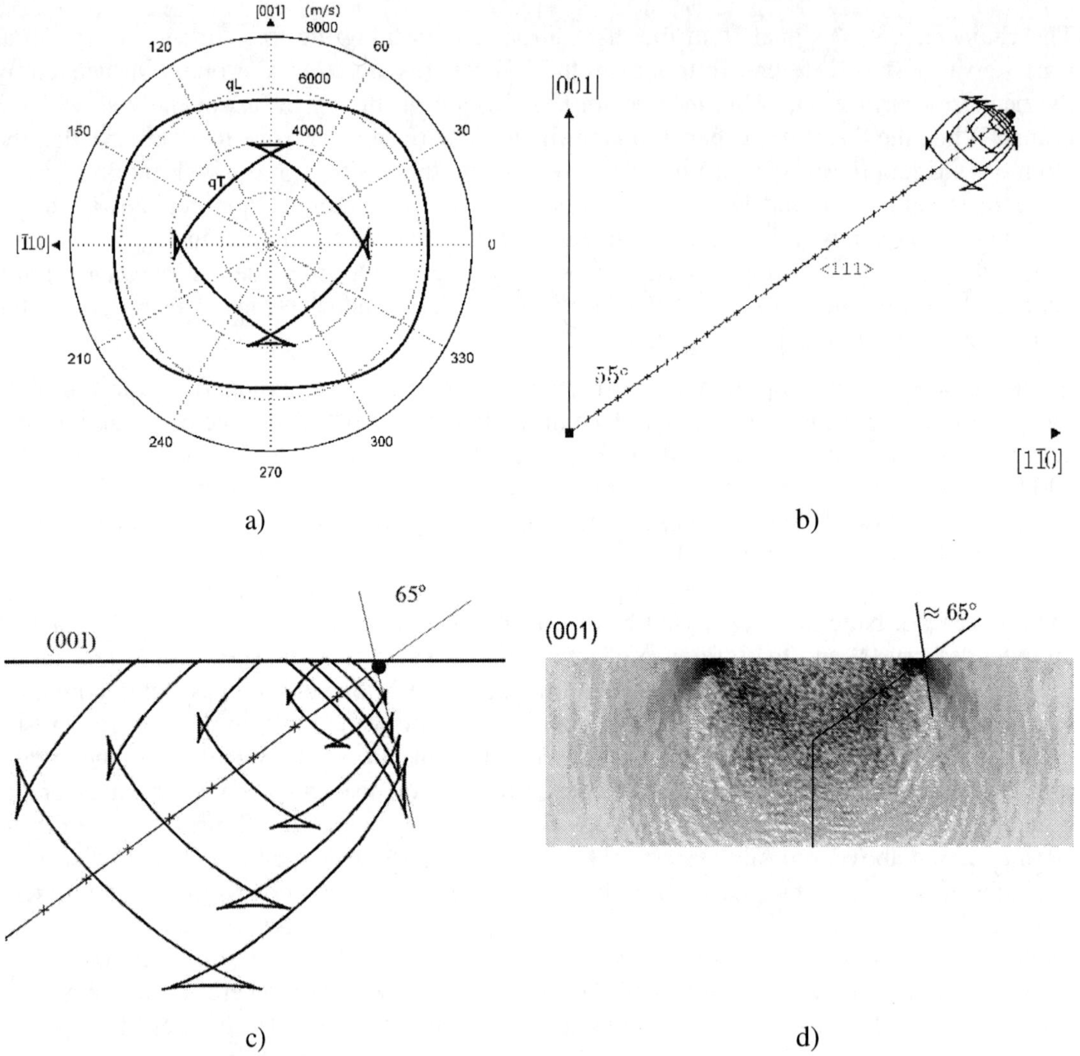

Figure 6. Reconstruction of the acoustic emission (AE) sources in anisotropic continuum of our crystal symmetry: a) section of the ray (wave) surfaces: group velocities of the quasi-longitudinal (qL) and quasi-transversal (qT) on (110) plane of bcc iron under plane strain (The group velocities will be calculated by means of `christoffel`, a Python tool for calculation direction-dependent phase velocities, polarization vectors, group velocities, power flow angels and enhancement factors based on the stiffness tensor of a solid [6]); b) AE patterns near by the final position (•) of the dislocation core, generated by the transonic dislocation, modelled as a moving source (+) of the qT-waves with a velocity $V_{\mathrm{disl}} = 1.36\ C_{\mathrm{T}}$. Initial position of the dislocation is identical with the crack tip point denoted via (■) and the moving positions (+) corresponds to Burgers vector in the slip direction <111> inclined under an angle of about 55° with respect to axis of the potential crack extension in the direction [001]; c) simplified reconstruction at the free surface created via a cut selection from Fig. 8b) near by the final position of the right dislocation core (•) approaching the free surface and Mach angle of about 65° corresponding to transonic velocity 1.36 C_{T}; d) comparison with a detail from MD for the transonic dislocation approaching the free (001) surface in Fig. 5b) (gray color scale in Matlab).

The qT waves are often denoted as the quasi-shear (qS) waves. It is well known that the slow (qS_1) and faster quasi-transversal qT waves (qS_2) must be recognized in anisotropic cubic crystals. For example: the velocity of the faster qT waves (presented in Fig. 8) in the <110> directions is given by the elastic constant C_{44}, while the velocity of the slower shear wave (qS_1) in these directions is given by $(C_{11}-C_{12})/2$. The waves have different polarization. The faster qT wave is polarized in the same plane where the direction of propagation lies, while the slower qT(qS_1) wave is polarized in a perpendicular plane.

In our crystal orientation (Fig. 1) the faster qT (qS_2) wave propagating in the [$\bar{1}$10] direction is polarized in the [001] direction, while the slower transversal (shear) wave (qS_1) is polarized in the perpendicular [110] direction. In Fig. 6 we do not consider the slower transversal wave qS_1, since we use plane strain conditions with zero displacements in the perpendicular [110] direction. These are relevant in the middle of the crystal (at *B*/2) for the edge dislocations emitted from the crack in the slip systems <111>{112}, since the Burgers vector lies on the observation plane (110), i.e. there is no component in the perpendicular direction [110].

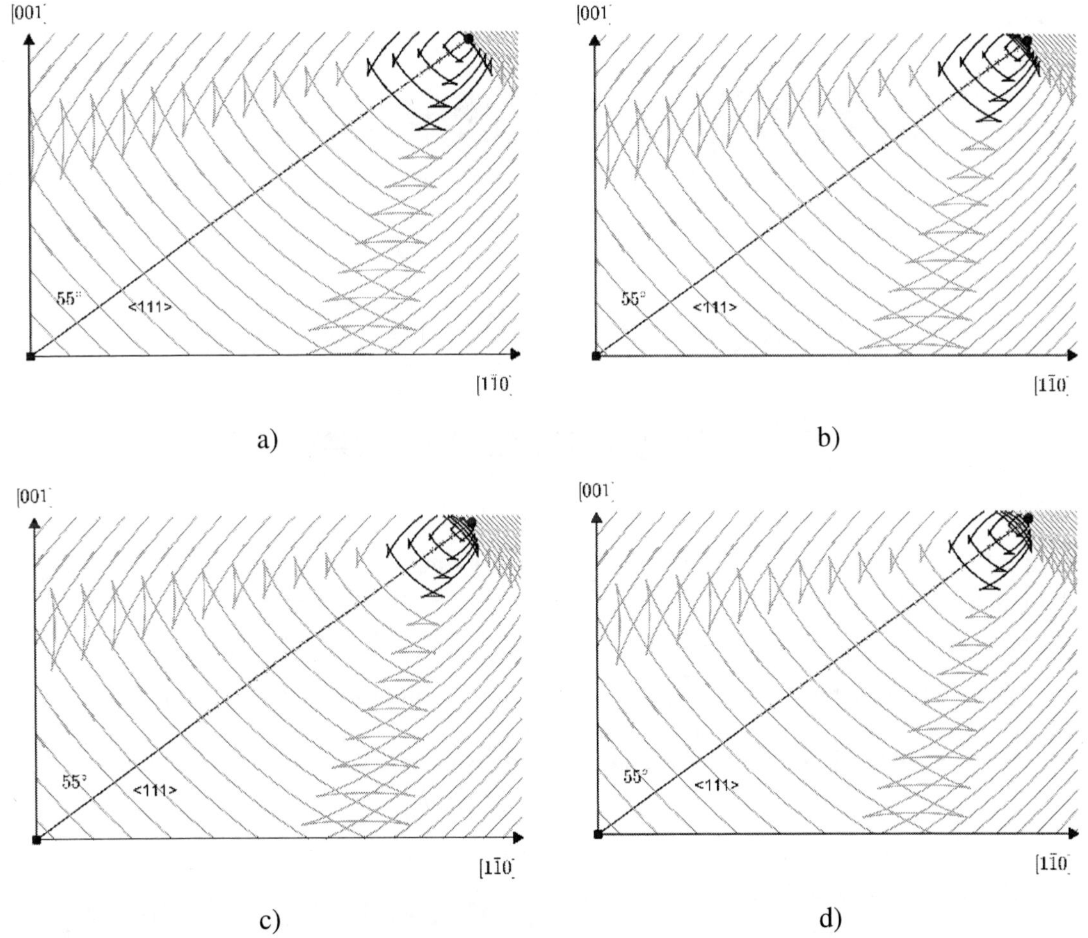

Figure 7. AE patterns near by the final position (•) of the dislocation core, generated by the transonic dislocation, modelled as a moving source (+) of the qT-waves with a velocity: a) $V_{disl} = 1.00\ C_T$, b) $V_{disl} = 1.2\ C_T$, c) $V_{disl} = 1.36\ C_T$ and d) $V_{disl} = 1.47\ C_T$.

Figure 7 shows AE patterns near by the final position (•) of the dislocation core, generated by the transonic dislocation (the black curves), modelled as a moving source (+) of the qT-waves with different velocities V_{disl}. The gray curves model the previous dislocation motion with the subsonic constant velocity $V_{disl} < 1\ C_T$ – see Fig. 4. Figure 7 illustrates that the different dislocation velocities V_{disl} lead to a different Mach angle between the envelope of the emitted shear waves and the direction of dislocation propagation <111>. At $V_{disl} = 1\ C_T$ it is roughly 90° (see Fig. 7a). With the increasing dislocation velocity above C_T (Figs 7b, c, d) this Mach angle decreases, i.e. it becomes sharper. It enabled us to identify more precisely the transonic dislocation velocity $V_{disl} = 1.36\ C_T = 4090$ m/s. This velocity lies closer to the lower limit in the broad interval <3724-7447> m/s following from the atomistic data presented in Fig. 4.

Note that arrival of the transonic dislocations to the narrow free BL surfaces (001) causes a fast jog formation on the free surface at time step ~11 900, which is a new AE signal that also propagates with a fast velocity 3400-3440 m/s above C_T. It can be recognized via the expanding wave patterns in Figure 8a) and b). Here, the black colour marks the highest $E_{kin}(li)$.

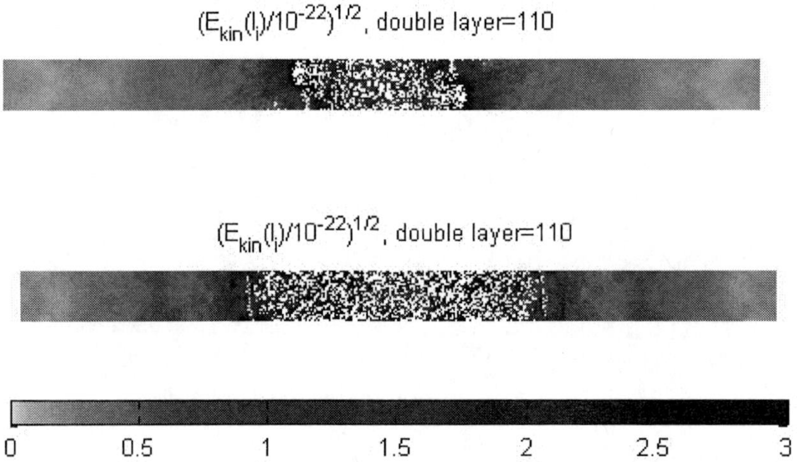

a)

b)

Figure 8. AE signal on *BL* free surface (001) via mapping of the local kinetic energies $E_{kin}(li)$ of the individual atoms: a) at time step 11 900; b) at time step 12 200.

The results presented in Fig. 8 agree with the continuum predictions from [7] that in anisotropic crystals of cubic symmetry, the surface waves may propagate in transonic regime, i.e. above the velocity 2474 m/s of the slower shear waves qS1 in the <110> directions.

4. Conclusions

Dislocation emission from edge crack $a/W = 0.8$ embedded in a 3D bcc iron crystal were studied via MD by a direct graphical treatment of the atomistic configurations, further via mapping of the local kinetic energies of individual atoms (revealing the acoustic emission from dislocations).

The graphical treatment of MD results show that the edge dislocation segments in the middle of the crystal after the emission move on the slip systems <111>{112} with a subsonic velocity up to a very close distance from the free surface (001) where they accelerate and penetrate the surface layers in transonic or supersonic regime.

An analysis of the acoustic emission from the moving dislocations has shown that dislocation motion in MD near by the free surface (001) can be accelerated to transonic regime (not supersonic), i.e. just above the velocity C_T of the transversal shear waves in the <111> direction.

Acknowledgements

The work was supported by the Institute Research Program RVO:61388998 and by the projects of the Czech Science Foundation GACR 17-22615S and GACR 17-12925S. The work of P. Hora was supported by the European Regional Development Fund under Grant No.CZ.02.1.01/0.0/0.0/15_003/0000493 (Centre of Excellence for Nonlinear Dynamic Behavior of Advanced Materials in Engineering).

References

1. P Hora, A Machová, J Červ and A Uhnáková, 'Kinetics of dislocation emission in 3D: from the crack toward the free surface', Kovove Mater., Vol 57, No 2, pp 75-93, 2019

2. J Červ, M Landa and A Machová, 'Transonic twinning from the crack tip', Scripta Materialia, Vol 43, No 5, pp 423-428, 2000.

3. JR Rice, 'Dislocation nucleation from a crack tip: An analysis based on the Peierls concept', J. Mech. Phys. Solids, Vol 40, No 2, pp 239-271, 1992.

4. GE Beltz and A Machová, 'Reconciliation of continuum and atomistic models for the ductile versus brittle response of iron', Modelling Simul. Mater. Sci. Eng., Vol 15, No 2, pp 65-83, 2007.

5. A Spielmanová, A Machová and P Hora, 'Crack-induced stress, dislocations and acoustic emission by 3-D atomistic simulations in bcc iron', Acta Materialia, Vol 57, No 14, pp 4065-4073, 2009.

6. JW Jaeken and S Cottenier, 'Solving the Christoffel equation: Phase and group velocities', Computer Physics Communications, Vol 207, pp 445-451, 2016.

7. N Favretto-Cristini, V Komaditich, J Garcione and F Cavallini, 'Elastic surface waves in crystals, Part 1: Review of the Physics', Ultrasonics, Vol 51, pp 653-660, 2011.

Optimization of Applied Magnetic Field Methods for Various Shape Defects by Eddy Current Testing Using Magneto Resistive Sensor

Takumi KOBARA, Yuto GODA, Hiroto SHOBU,
Kenji SAKAI, Toshihiko KIWA, Keiji TSUKADA

[1] Okayama University, 3-1-1 Tsushimanaka, Kita-ku, Okayama, Japan
E-mail: py0u6os5@s.okayama-u.ac.jp

Abstract
In this study, we investigated the optimization of applied magnetic field methods for defect detection for various shapes of defects made of magnetic and non-magnetic materials using the eddy current testing with a highly sensitive magnetic sensor. From the experiment and simulation results, it was found that the change of the magnetic field distribution caused by the eddy current increased when the sensor was placed directly above the induction coil of the wire-type method. Moreover, in the solenoid coil system, it was found that the detection sensitivity increased when the magnetic sensor was placed close to one side of the coil compared to that placed at the center. In addition, it was found that the signal change was different in the electromagnetic characteristics of the measurement sample. For the non-magnetic material, as the frequency increased, the signal to noise ratio (SNR) of the detected defect signal increased because the signal was generated from only the eddy current. On the other hand, for the magnetic material, there was an optimum frequency for detecting the defect because the signal was generated due to both the magnetization and the eddy current. As a results, the optimum applying method of magnetic field for the eddy current testing using magnetic sensor was clarified.

Keywords: Nondestructive testing, Eddy current testing, Magnetic material, Non-magnetic material, Magneto resistive sensor

1. Introduction

Micro defects occur on the surface or inside metal materials during their production, processing for a metal structure, or caused by the aged deterioration. Therefore, a non-destructive testing that can inspect and evaluate the micro defect of early stage is required. There are many types of non-destructive inspection methods, such as eddy current testing [1-3], magnetic flux leakage testing [4-6], and radiography testing. The eddy current testing method has many advantages such as non-contact, high speed, automation, and can be applied to conductive materials. In the conventional eddy current testing, a combination of induction and detection coils is widely used, however, the magnetic sensor which has high sensitivity and spatial resolution even at low frequencies is recently used instead of the detection coil. In the previous study, we have attempted to detect the micro defects by the eddy current testing using magnetic sensor. As a result, although the signal change due to the micro defect was confirmed, there was a problem that the spatial resolution and signal to noise SNR (SNR) was low when a general solenoid coil was used.

When the solenoid coil was used for the detection coil, the magnetic response from the defects was buried in the applied magnetic field because the ratio of the surface area formed by the induction coil is small relative to the size of micro defects. Consequently, the SNR of the detected defect signal was considered to be low. Therefore, in this study, in order to improve SNR of detecting micro defects, various application methods of magnetic field, such as a solenoid or single wire type, were compared and the improvement in the detection performance of micro defects was investigated.

2. Measurement system

Figure 1(a) shows photographs of three applied magnetic field methods, and Fig. 1(b) shows the pattern diagram of the sensor position relative to the induction coil. In Fig. 1(a), the positional relationship between the induction coil and the sensor is as follows. In the left photograph, the sensor was placed at the center of the coil. In the center photograph, the sensor was placed one side of the coil (lower side of the photo) [7]. In the right photograph, a sensor was placed in the center of the induction wire. Hereafter, these are called solenoid (sensor center), solenoid (sensor edge), and wire type, respectively. The inner and outer sizes of the solenoid (sensor center) were 2.3 × 2.3 mm^2 and 3.8 × 3.8 mm^2, and the turn number was 32 turns. The inner and outer sizes of the solenoid (sensor edge) were 6.0 × 2.5 mm^2, and 7.7 × 4.5 mm^2, and the turn number was 60 turns. The wire type consisted of a single turned coil.

Figure 2 shows the measurement system. The measurement system consisted of an oscillator that generates an alternating current (AC) signal, three types of coils for applying an AC magnetic field to the test sample (aluminium and steel), a differential amplifier for amplifying the signal from the defects, a tunnel magneto resistive (TMR) sensor for detecting change of the defect signal by the eddy current testing, a lock-in amplifier for lock-in detection of the difference between the defect signal acquired by the TMR sensor and the reference signal from the oscillator, a PC acquiring the data of lock-in detected signal, a X-Y automatic scanning stage to drive the TMR sensor.

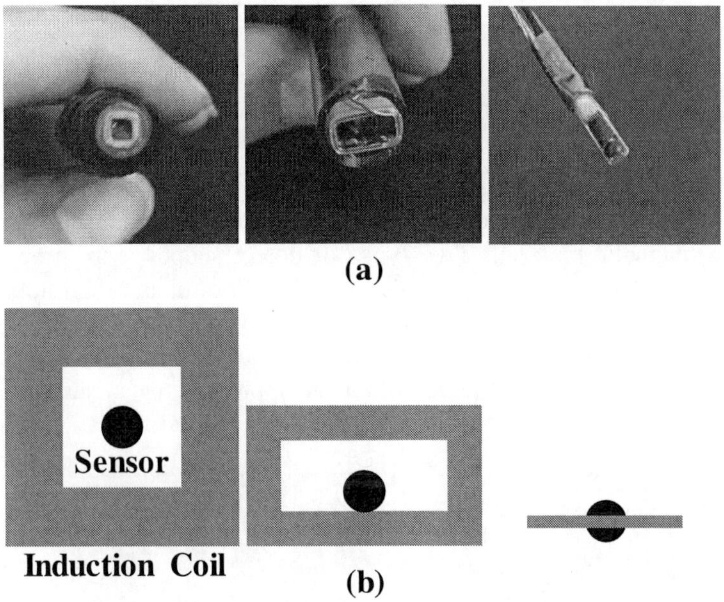

Fig. 1. Applied magnetic field methods (from the left: solenoid (sensor center), solenoid (sensor end), wire type): (a) Photograph and (b) Pattern diagram.

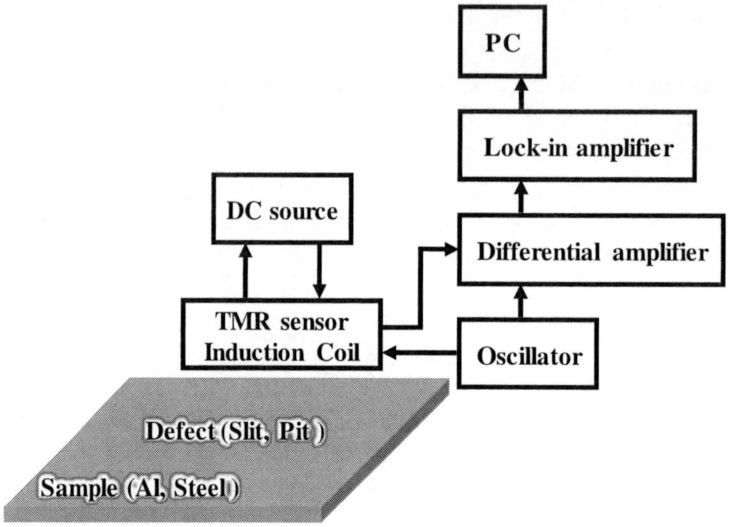

Fig. 2. Measurement system.

3. Experimental result

3.1 Measurement sample

Figure 3 shows the measurement sample diagram. Two types of measurement samples were prepared: one is an aluminium plate which is a non-magnetic material, the other is a steel plate which is a magnetic material. Two types of defect shapes were prepared: pit and slit. The diameter of pit defect was 2.0 mm, the depth was 1.0 mm, and the thickness was 10 mm. The length of slit defect was 4.0 mm, the width was 0.4 mm, and the thickness was 20 mm. The measurement conditions are as follows, the frequency was 50 kHz, the line scanning was performed at 0.1 mm intervals, the lift-off was appressed to a sample, the number of measurements was 81 points (slit defect), and 121 points (pit defect).

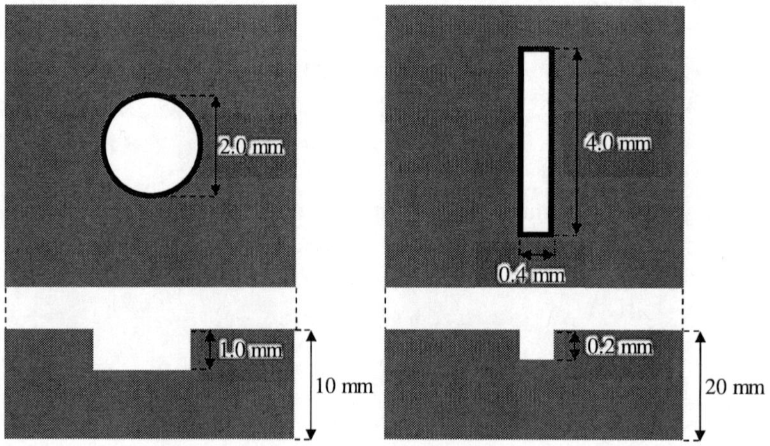

Fig. 3. Measurement sample.

3.2 Line scanning at pit defect

Figure 4 shows the signal intensity of single line scanning in Al plate of non-magnetic material using various applied magnetic field methods, and Fig. 5 shows the signal intensity when the measurement sample was a steel plate of magnetic material. The direction of the detected magnetic field was vertical to the surface of measurement sample.

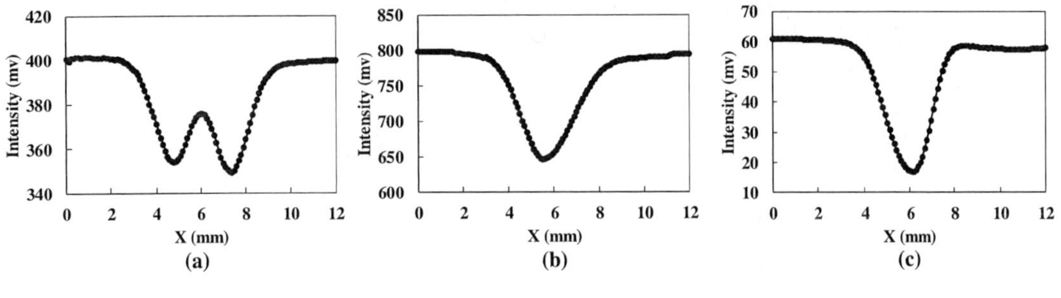

Fig. 4. The signal intensity of Al (a) solenoid (sensor center), (b) solenoid (sensor end), (c) wire type.

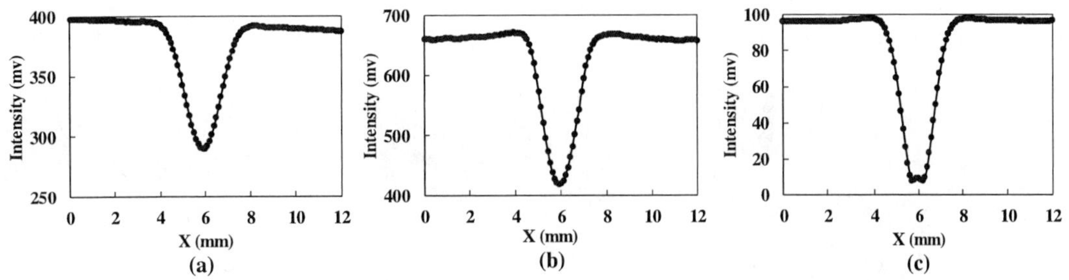

Fig. 5. The signal intensity of steel (a) solenoid (sensor center), (b) solenoid (sensor end), (c) wire type.

From both results shown in Figs. 4 and 5, when the solenoid (sensor center) was used, two peaks was observed at the defect position for the non-magnetic material, and a single peak was observed at the defect position for the magnetic material. This shows that two peaks of Al were caused by the concentrated eddy currents at the edge of the defect. On the other hand, in the steel materials, a single peak was observed because the distribution of the eddy current was uniformly distributed near the defect and the signal intensity decreased at the center of the defect. However, as seen in Figs 4(b), (c) and Figs. 5(b) and (c), it was confirmed that a single peak was observed in both materials. In addition, the signal change was constant at the defect center, as shown in Fig. 5(c). This is because the bottom surface of the defect is flat, and the magnetic response generated at the bottom surface becomes constant if the spatial resolution is high. This result indicates that the wire type has higher spatial resolution than the other applied magnetic field methods.

To clarify the difference of three applied magnetic field method, the simulation of the distribution of the eddy current and magnetic field was conducted by the finite element method for two materials. The model used for calculation was the pit defect measured in Figs. 4 and 5. For the simulation, a commercially available software of JMAG-Designer was used. Figure 6 shows the pattern diagram of the positional relationship between the defect and the induction coil used for the simulation. Figure 7 shows the calculated signal intensity of the eddy current distribution of Al material when the one side of each coil was located just above the defect center. Figure 8 shows the calculated magnetic field distribution generated by the eddy currents. The values of mapping show the sum of the magnetic field intensity in the X, Y, and Z axis directions. The eddy current and magnetic field distribution when the sample was a steel were shown in Figs. 9 and 10, respectively.

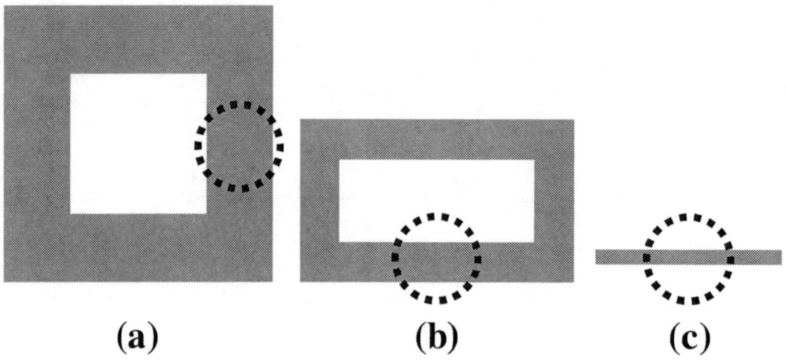

Fig. 6. The pattern diagram of the positional relationship between the defect and induction coil for thesimulation (a) solenoid (sensor center), (b) solenoid (sensor end), (c) wire type.

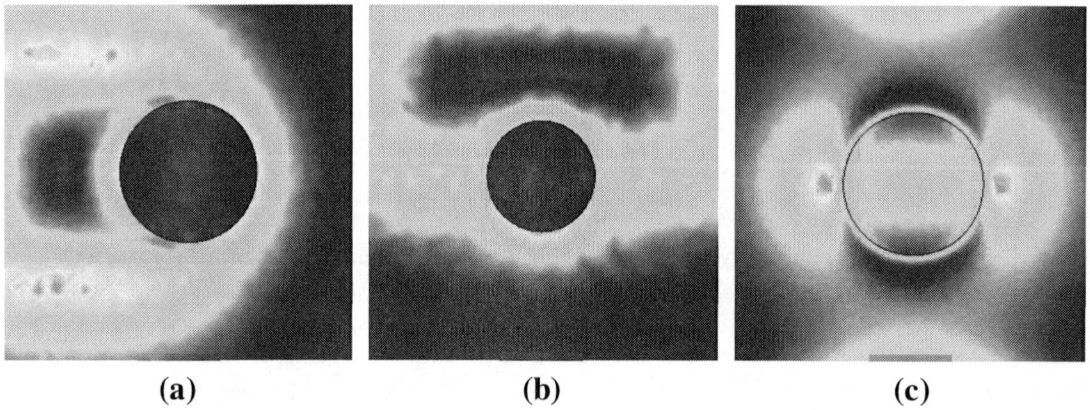

Fig. 7. Eddy current distribution of Al (a) solenoid (sensor center), (b) solenoid (sensor end), and (c) wire type.

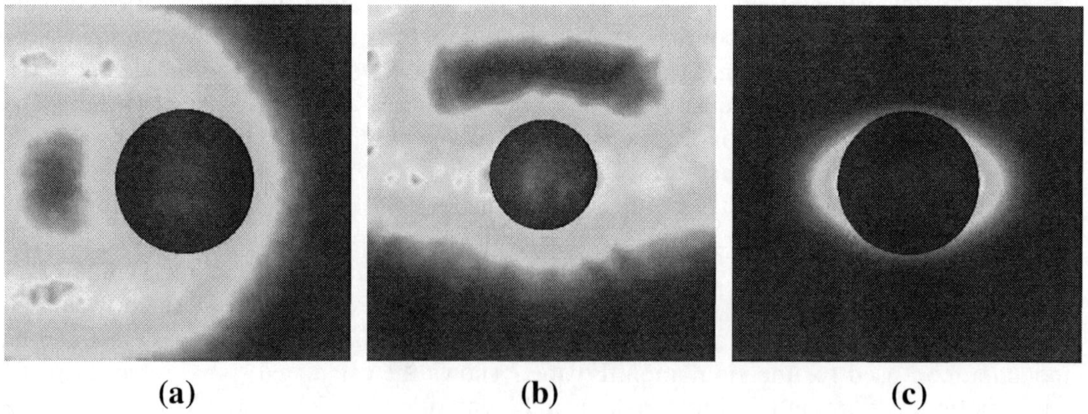

Fig. 8. Magnetic field distribution of steel (a) solenoid (sensor center), (b) solenoid (sensor end), and (c) wire type.

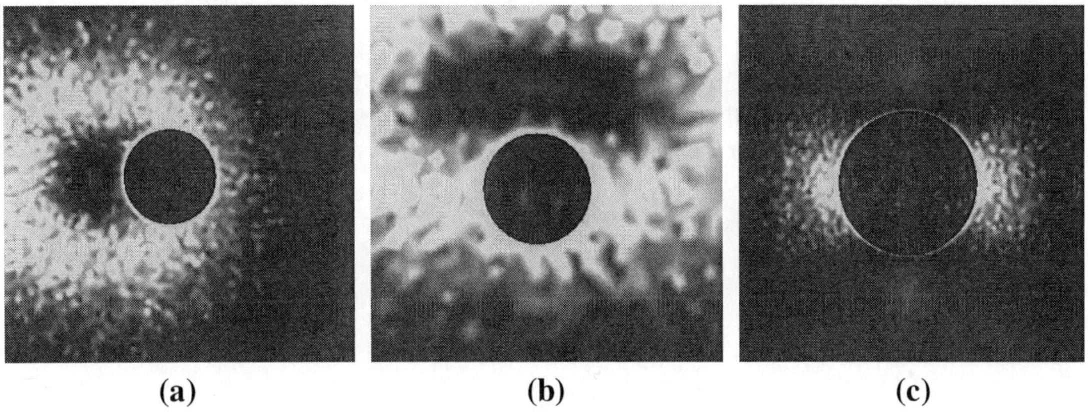

Fig. 9. Eddy current distribution of Steel (a) solenoid (sensor center), (b) solenoid (sensor end), and (c) wire type.

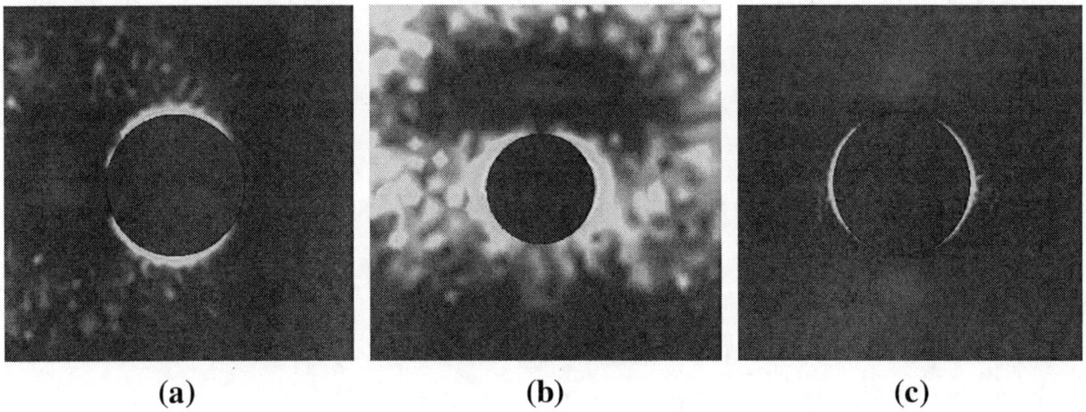

Fig. 10. Magnetic field distribution of steel (a) solenoid (sensor center), (b) solenoid (sensor end), and (c) wire type.

It was confirmed from Fig. 7(a) that the eddy current concentrated at the edge of the defect was not uniform. On the other hand, as seen in Figs. 7(b) and (c), the eddy current was distributed symmetrically at the edge of the defect. Figure 8 indicate that the magnetic field was concentrated at the edge of the defect only when the case of Fig. 8(c). When the sample was a steel plate, the concentration of the eddy currents at the edge of the defect was not be observed, as shown in Fig. 9. On the other hand, as shown in Fig. 10, the concentration of the magnetic field at the edge of the defect was observed in all cases. From these results, it was found that Al material with high electrical conductivity induces much eddy currents than steel material, and the steel material with high permeability shows the concentration of magnetic field distribution which is larger than that of Al material.

The ratio of signal change due to the pit defect with respect to the non-defect part for various applied magnetic field methods is shown in Fig. 11.

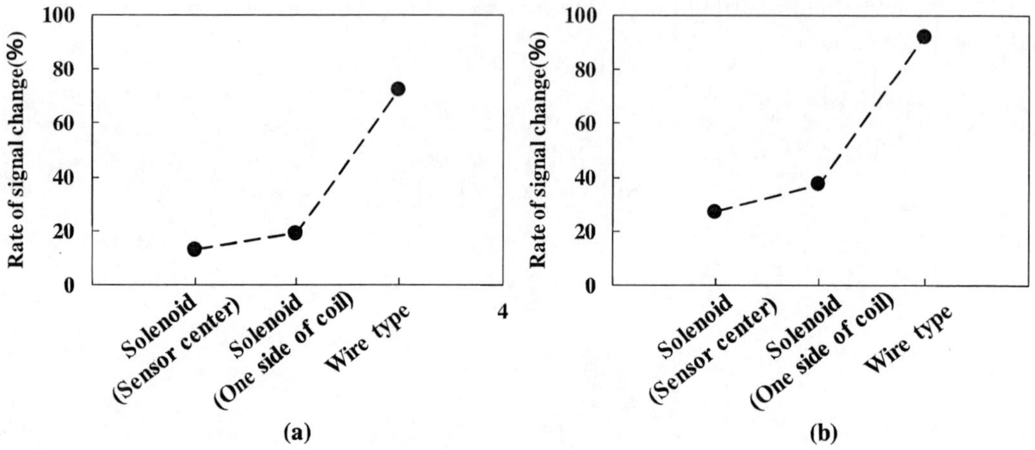

Fig. 11. The ratio of signal change (a) Al (b) Steel.

From the result of Fig. 11, it was found that the signal change increased by placing a sensor on one side of the coil when a solenoid type used. Further, it was found that the wire type showed the largest signal change when detecting micro defects compared to the other applied magnetic field methods.

3.3 Line scanning at slit defect

Figure 12 shows the signal intensity of single line scanning in Al material of non-magnetic material using various applied magnetic field methods, and Fig. 13 shows that in the case of steel material.

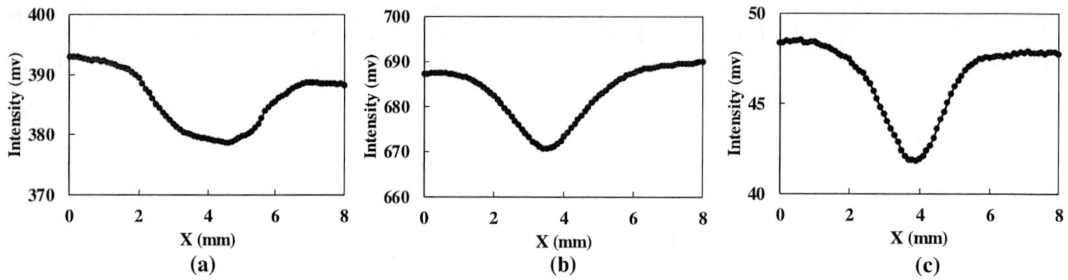

Fig. 12. The signal intensity of Al (a) solenoid (sensor center), (b) solenoid (sensor end), and (c) wire type.

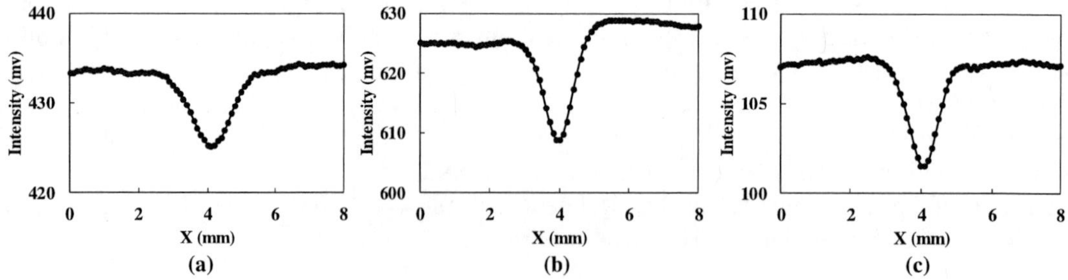

Fig. 13. The signal intensity of Steel (a) solenoid (sensor center), (b) solenoid (sensor end), and (c) wire type.

From the both results of Figs. 12. and 13, it was confirmed that a slit defect has a single peak in the center of the defect regardless of the non-magnetic and magnetic materials. This is because the eddy current flows in the slit defect so that the eddy current distributes underside of the defect regardless of the material of Al or steel. In addition, when the change of the signal intensity of Al measured by each applied magnetic field method was compared, the signal change of the wire type at the defect was remarkably observed compared to that using the solenoid type. Similarly, in the case of steel materials, it was remarkably observed that the wire type could suppress the drift caused by the magnetization and the signal change at the defect was strong compared to the other applied magnetic field methods. From this result, in the slit defect, it was found that the wire type has the high spatial resolution which is similar to the case of pit defect.
Figure 14 shows the ratio of signal change due to the slit defect with respect to the area without defect using various applied magnetic field methods for each material.

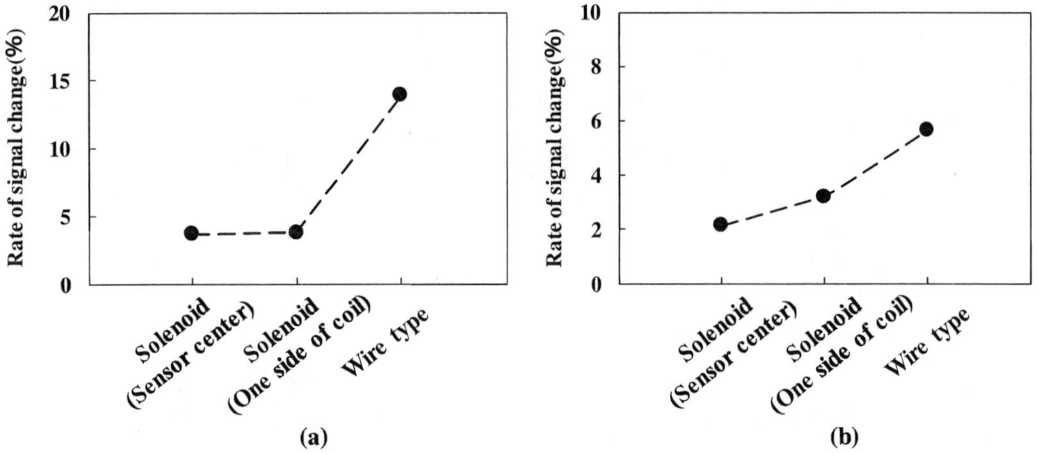

Fig. 14. The ratio of signal change (a) Al (b) Steel.

It was found from Fig. 14 that the wire type had the largest signal change for micro defects compared to the applied magnetic field methods of solenoid type.

4. Conclusion

In this study, three types of applied magnetic field methods was used to detect micro defects and the detection performance was investigated for various defect shapes made of magnetic and non-magnetic material. As a results, it was shown that the signal change of the wire type was the largest. In addition, it was found that the wire type had high spatial resolution compared to the other applied magnetic field methods and the possibility of detecting smaller micro defects was indicated.

References

1. K. Tsukada, T. Kiwa, T. Kawata. Y. Ishihara, "Low-Frequency Eddy Current Imaging Using MR Sensor Detecting Tangential Magnetic Field Components for Nondestructive Evaluation", IEEE Trans. Magn., vol. 42, no 10, pp. 3315-3317, 2006.

2. R. Ghoni, M. Dollah, A. Sulaiman, F. M. Ibrahim, "Defect Characterization Based on Eddy Current Technique: Technical Review', Adv. Mech. Eng., Vol. 2014, pp 1-11.
3. J. G. ía-Martín, J. G. ómez-Gi, E. V. ázquez-Sánchez, "Non-Destructive Techniques Based on Eddy Current Testing', Sensors, vol. 11, pp. 2525-2565, 2011.
4. K. Tsukada, M. Yoshioka, Y. Kawasaki, T. Kiwa, "Detection of back-side pit on a ferrous plate by magnetic flux leakage method with analyzing magnetic field vector", NDT&E Int., vol. 43, pp. 323-328, 2010.
5. Y. Nakamura, Y. Majima, K. Sakai, T. Kiwa, K. Tsukada, "Optimization of magnetic-field component detection for unsaturated AC magnetic-flux-leakage testing to detect cracks in steel", APCNDT 2017, 2017.
6. Y. Sun, Y. Kang, "Magnetic mechanisms of flux leakage non-destructive testing", Appl. Phys. Lett., vol. 103, 184104-1-5, 2013.
7. M. Hayashi, T. Saito, Y. Nakamura, K. Sakai, T. Kiwa, I. Tanikura, K. Tsukada, "Extraction Method of Crack Signal for Inspection of Complicated Steel Structures Using A Dual-Channel Magnetic Sensor", Sensors vol. 19, 30012019, 2019.

Influence of pre-magnetization and demagnetization of a steel wire for its diagnostic signal

Paweł MAZUREK[1], Mirosław WITOŚ[2], Jerzy KWAŚNIEWSKI[1], Maciej ROSKOSZ[1]

[1]*AGH University of Science and Technology, Mickiewicza 30, 30-059 Cracow, Poland*
[2]*Air Force Institute of Technology, Ksiecia Bolesława 6, 01-494 Warsaw, Poland*
e-mail: pmazurek@agh.edu.pl

Abstract
The article focuses on the very process of magnetization and demagnetization. As far as the measurement is concerned, it is presented as follows. Three identical, new steel wire samples from the production line, were subjected to the processes: one of them was magnetized with a strong magnetic field (like in a active method), the other was subjected to a demagnetization process (passive method), while the third was left unchanged. A diagnostic signal was recorded for each of these samples using the SpinMeter-3D sensor. The signal was measured in the three axes of the Cartesian system. Then, each of these samples was subjected to two-way bending until breaking. Diagnostic signals for each of these samples were registered again and next obtained results were compared very closely.

Keywords: technical diagnostics, steel wire, pre-magnetization, demagnetization, magneto-mechanical effects

1. Introduction

Steel wires have many different applications and are used in almost every area of industry. As a result of continuous use, they inevitably have been degraded. Most often, this degradation is caused by four phenomena: fatigue, friction fatigue, wear or corrosion. These phenomena may occur on their own, but almost always occur in all combinations and with different intensities. The basic forms of fatigue of steel wires are: tension-tension fatigue, free-bending fatigue, torsion fatigue and bending over sheaves fatigue [1].
Sudden rupture of the rope can threaten the life and health of people and cause costly downtime. As a result, accidents resulting from damage to wire ropes are a serious problem and a challenge for designers and constructors - especially when it comes to the safety of human life. The basis for increasing safety and improving the efficiency of rope use is a good knowledge of the degradation process. The fatigue processes mentioned above have the greatest impact on the development of degradation [2].
Values of a residual magnetic induction change as a result of the work of a steel wire, i.e. its bending, twisting or stretching. The value of the induced magnetic field in the wire is also affected by the damage occurring. Under the influence of cyclically changing work loads, due to the effects of magnetomechanics, thermomechanics, changes in electromagnetic properties occur in the wires. Anomalies of the residual magnetic field are used to identify imperfections of steel wires such as deformations, wear and corrosion. The most common method of non-destructive testing of steel wires is the active magnetic method, which uses imperfections leaked by a strong, formatted by an external source, magnetic field. In this case, the steel wire rope is magnetised only before or during the test, thus losing important diagnostic information. An increasingly popular method using a residual magnetic field leakage does not require pre-magnetization. Magnetic flux leakage (MFL) detection has the advantages of accurate positioning obvious target defect. The MFL information on the surface of the wire can be converted into magnetic images so that the circumferential and axial position of the wire defects can be displayed more intuitively. By fusing the visible and magnetic images, it is possible to use of their information and the defect recognition rate can be improved [3, 4, 5, 6].

2. Methods of diagnostic steel wires

Magnetic diagnostic methods may be divided into passive and active ones. The use of external magnetic field acting against a tested component is a common feature of active magnetic diagnostic methods and the evaluation of object's technical condition is made by analysing recorded magnetic parameters. The phenomenon of anomalies existing in distributions of magnetic field intensity components over a tested surface is used for magnetic detection of inhomogeneities within a tested object. The magnetic field flux passing through the material is dispersed in a place of inhomogeneity which produces magnetic flux leakage outside the tested object because on the edges of discontinuity the specific "magnetic poles" are created, which in turn generates magnetic field above this discontinuity [7].

2.1 The mechanism of Magnetic Memory Method

The idea of MMM is the magneto-mechanical coupling of ferromagnetic component under weak magnetic field (like geomagnetic field). For ferromagnetic material, the total free energy of magneto-crystalline under no stress state can be expressed by Eq. (1).

$$\boldsymbol{E = E_k + E_{ms} + E_{el} + E_n + E_d} \tag{1}$$

where $\boldsymbol{E_k}$ is magneto-crystalline anisotropy energy; $\boldsymbol{E_{ms}}$ is magneto-elastic energy; $\boldsymbol{E_{el}}$ is elastic energy; $\boldsymbol{E_n}$ is external magnetic field energy; $\boldsymbol{E_d}$ is demagnetizing energy

When subjected to external force, the crystal of ferromagnetic component generates greater stress and strain energy. The total free energy in crystal at this point can be referred to Eq. (2).

$$\boldsymbol{E = E_k + E_{ms} + E_{el} + E_n + E_d + E_\sigma} \tag{2}$$

where $\boldsymbol{E_\sigma}$ is the stress energy of materials under external force.

According to Ref. [8], when the ferromagnetic component subjects to stress σ and the material is assumed as magnetostrictive isotropy, the internal stress energy E_σ can be presented in Eq. (3).

$$\boldsymbol{E_\sigma = -\frac{3}{2}\sigma\lambda_\sigma \cos^2\theta} \tag{3}$$

where $\boldsymbol{\lambda_\sigma}$ is magnetostrictive coefficient of material; $\boldsymbol{\theta}$ is the subtended angle between stress direction and magnetization direction

Not only can stress increase elastic energy and magneto-elastic energy of ferromagnetic component, but change magnetization direction of domain walls as per magneto-elastic effect and magneto-mechanical effect. With the completion of domain directional movement, such irreversible magnetization vector will still exist. Consequently, the SMFL signals on component surface will also exhibit some regularity. Under the stress and geomagnetic field, the directional movement of domain and magnetostriction effect in ferromagnetic component will cause inconsecutive distribution of macro-magnetic properties in stress concentration zone (SCZ). This status is still retained even after the stress is released. It is called magnetic memory effect.

Magnetization of the ferromagnetic material can be provided by 3 independent variables:

- external magnetic field with intensity **H**,
- stress state σ_{ij}, which is associated with applications based on magneto-mechanical effects,
- material temperature changes ΔT.

The resultant non-linear influence of 3 models on magnetization attributes the Eq. (4)

$$\mathbf{M} = \mathbf{M}_i + \mathbf{M}_r = f(\mathbf{H}, \sigma_{ij}, \Delta T) = (1 + k_H)(1 + k_\sigma)(1 + k_{\Delta T})\mathbf{M}_0 \quad (4)$$

where M_i- magnetization induced by operational operations; M_r – residual magnetization present in the material after loading; M_0 – initial magnetization of the element; $k_H, k_\sigma, k_{\Delta T}$ – functions affecting magnetic fields, stress and temperature changes

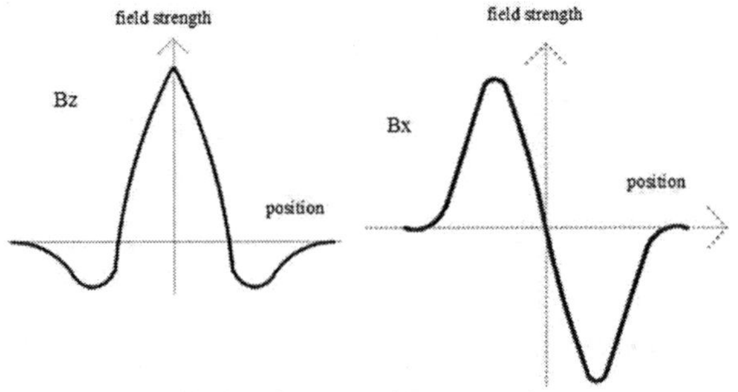

Fig. 1. Theoretical field distribution for Bz and Bx around a stress-concentration zone [9]

Theoretical field distribution for Bz and Bx around a stress-concentration zone is shown on the Fig. 1.

2.2. The active magnetic method

As a classical NDT method, the principle of the MFL technique is relatively simple. That is, when a strong magnetic field is applied to a ferromagnetic material, any geometrical discontinuity in the test object will cause the field to leak out of the object into the air. The flux leakage can be monitored by a magnetic field sensor and used to estimate the dimensions of the defect. Although the MFL phenomenon is easily understood, the design and analysis of MFL systems involve complicated interactions between the excitation field, leakage flux and the defects in the material. There are several important aspects to be considered. Firstly, the level of the excitation magnetic flux should be large and homogenous to allow the magnetic flux variation to occur at the location of a defect; secondly, the sensors should be located close to the position at which the changes in the magnetic field density, originating from the defect, are distinct from the background noise; additionally, developing an effective inversion method to identify the defective characters by the recorded MFL signals is difficult since the defect is irregular. [10, 11, 12, 13]

3. Effects on the wire

3.1. Magneto-mechanical effects

The magneto-mechanical effect is a fundamental feature of ferromagnetism. The fact that the application of external stresses alters the flux density of a magnetised ferromagnet, and thus the shape, and size of its hysteresis loops is easily verifiable. It is even possible to develop empirical equations to predict these change. The responses of different magnetic materials to stress are well documented. The relationship between the magneto-mechanical effect and magnetostriction means it is possible to quite accurately determine the degree and direction of changes induced by stress. However, a conceptual theory as to the actual mechanism by which stress and magnetism interact has been more elusive. Biaxial strain and magnetisation data have been obtained from an annealed mild steel plate cut to the shape of a cruciform, and the resultant curves have been analysed with the intention of developing a theory which explains the forms and features of the curves rather than the production of a numerical model [14, 15, 16, 17].

3.2. Thermoelastic effects

Thermoelectric effects can generate magnetic fields by electric current circulation that, in the presence of sensible measuring apparatus, can hinder performance and even lead to misleading data. Although well-known, the thermoelectric effects are not normally accounted during the design phase of electrical and/or electronical systems due to its relatively small impact [18].

3. Details of the measurement

Three identical, new steel wire samples 600mm length, diameter 3mm from the production line, were subjected to the processes: one of them was magnetized with a strong magnetic field (like in a active method), the other was subjected to a demagnetization process (passive method), while the third was left unchanged. A diagnostic signal was recorded for each of these samples using the SpinMeter-3D sensor [Fig. 2, 3, 4].

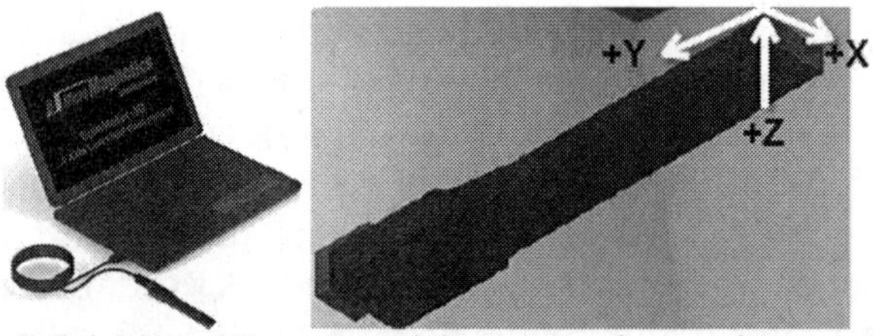

Fig. 2. SpinMeter-3D sensor and the system of measuring axes [19]

The technical specification of the SpinMeter 3D was shown in the Table 1. [19]

Table 1. Technical specification of SpinMeter 3D [19]

Parametr	SpinMeter 3D
Producent	Micromagnetic
Type	3D MTJ/TMR
Full scale range	±1000 μT
Sensitivity	0,10 μT
Noise	0,25 μT rms
Data rate	80 Hz

Fig. 3. Equipment and procedure of bending steel wire

Fig. 4. Procedure of bending steel wire - details

The signal was measured in the three axes of the Cartesian system. Then, each of these samples was subjected, according to the PN-ISO 7801:1996 norm, to two-way bending until breaking. Diagnostic signals for each of these samples were registered again and next obtained results were compared very closely.

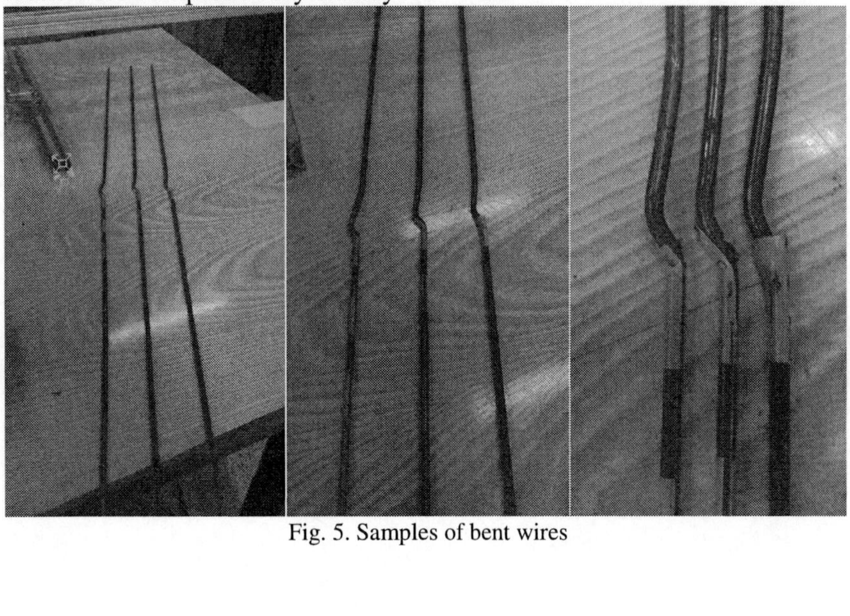

Fig. 5. Samples of bent wires

4. Results

The measurement results obtained are presented in the following charts. Fig.6. presents the number of bend of each sample. Figures 7-9 present the state magnetization in each sample.

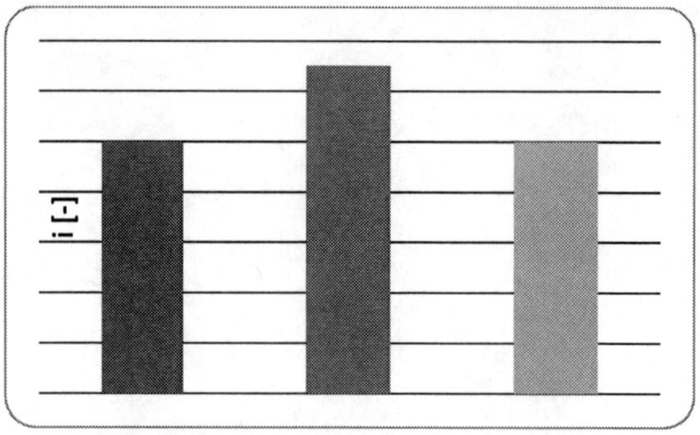

Fig. 6. Number of bends of each sample

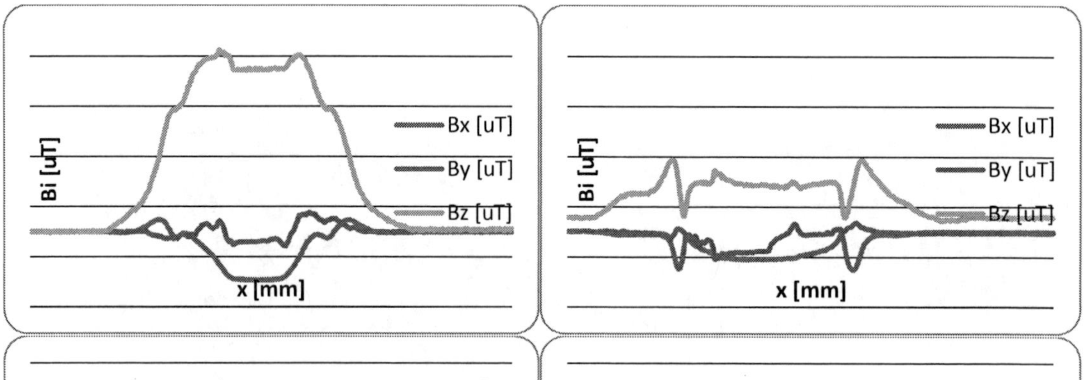

Fig. 7. Sample 1 – before and after fracture

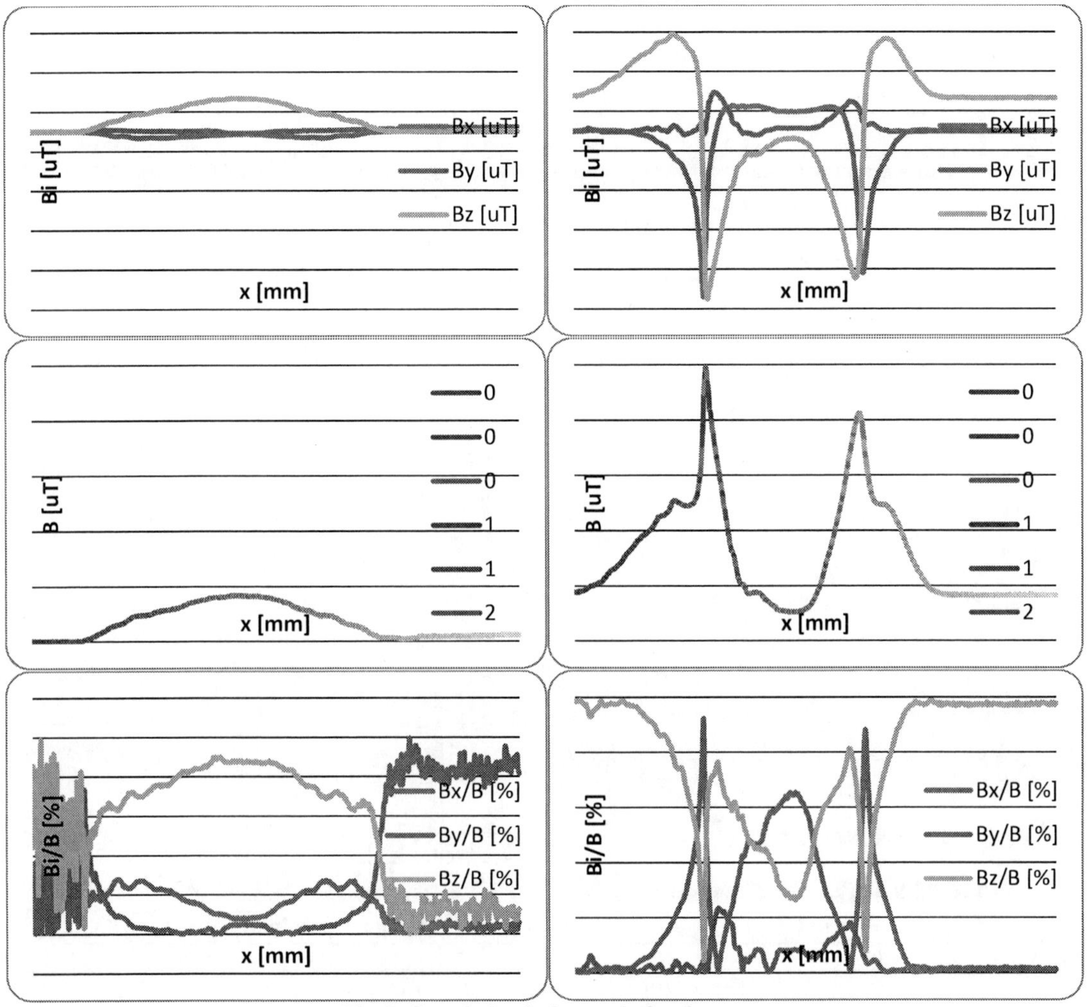

Fig. 8. Sample 2 – before and after fracture

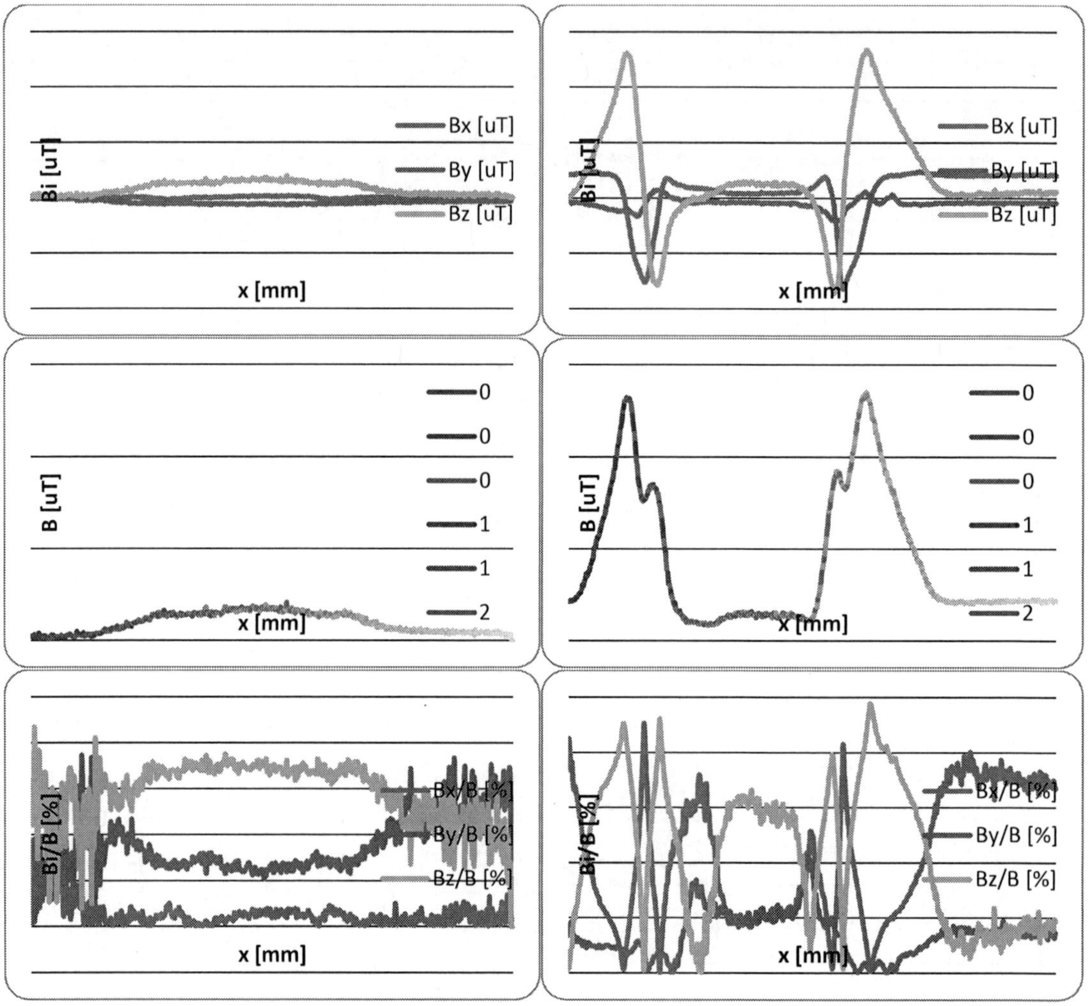

Fig. 9. Sample 3 – before and after fracture

The graphs show noticeable magnetic changes for each sample. In any case, the magnetic signal has changed after the wire was broken. This change, however, varies depending on the process the sample was subjected to. The largest diagnostic signal has a pre-magnetized sample. This is the change with the highest magnetic induction amplitude of 1200 uT. A visible change in the level of magnetization also appears in the case of initial demagnetization - in this case, however, the signal amplitude is of the order of 200uT. Very little visible change occurred in the case of wire that was not initially subjected to any process.

5. Conclusions

Both initial magnetization and demagnetization affect the diagnostic signal - it is more visible. The pre-magnetized wire was subjected to as many as 13 bends before refraction, unlike the other two, which only withstand 10 bends. This allows us to suppose that magnetization strengthens the material.

The presented graphs show that the most of the diagnostic signal is transferred by the z component. The x component also informs about the change, however, the change in the y component results from the inaccuracy of the sample and the inaccuracy of the measurement.

Initial magnetization of the sample - before the damage occurs, affects the diagnostic signal exposure. The damage is much more visible. The same happens with demagnetization. These processes carried out before the damage occurred allow the diagnostic signal to be displayed. However, in the case of passive methods, strong magnetization of the material may disturb subsequent test results. In the case of the MMM method, after the test, magnetostructions appear in the sample only from external forcing - including stress or damage.

References

1. Mouradi, H., El Barkany, A., & El Biyaali, A. (2018). Steel wire ropes failure analysis: Experimental study. Engineering Failure Analysis, 91, 234–242. doi:10.1016/j.engfailanal.2018.04.019
2. Van der Horst, M. P., van Kreveld, S. L., & Kaminski, M. L. (2018). Effect of stress-induced magnetization on crack monitoring by self magnetic flux leakage method. International Journal of Applied Electromagnetics and Mechanics, 1–18. doi:10.3233/jae-180031
3. Erena D, Vázquez Valeo J, Navarro C, Domínguez J. Fatigue and fracture analysis of a seven-wire stainless steel strand under axial and bending loads. Fatigue Fract Eng Mater Struct. 2019;1–13. doi.org/10.1111/ffe.13096
4. Ono, K. (2019). Size Effects of High Strength Steel Wires. Metals, 9(2), 240. doi:10.3390/met9020240
5. Onur, Y. A., İmrak, C. E., & Onur, T. Ö. (2019). Discarding lifetime investigation of a rotation resistant rope subjected to bending over sheave fatigue. Measurement, 142, 163–169. doi:10.1016/j.measurement.2019.04.078
6. Y Deffo, P Tsafack, Benjamin Ducharne, M. Le, P. Cottinet. Effect of bending stress on the magnetic properties of electrical steel using needle probe method. COMPUMAG 2019, Jul 2019, PARIS, France.
7. Dybała J, Nadulicz KM. Diagnostics of welded joints using passive and active magnetic methods. Diagnostyka. 2017;18(4):51-59
8. Z.D. Wang, K. Yao, B. Deng and K.Q. Ding, Quantitative study of metal magnetic memory signal versus local stress concentration, *NDT & E International* 43(6) (2010), 513–518.
9. Ma, X., Su, S., Wang, W., Yang, Y., Yi, S., & Zhao, X. (2019). Damage location and numerical simulation for steel wire under torsion based on magnetic memory method. International Journal of Applied Electromagnetics and Mechanics, 1–24. doi:10.3233/jae-180075
10. Su, S., Zhao, X., Wang, W., & Zhang, X. (2019). Metal Magnetic Memory Inspection of Q345 Steel Specimens with Butt Weld in Tensile and Bending Test. Journal of Nondestructive Evaluation, 38(3). doi:10.1007/s10921-019-0603-8
11. Xie, Z., Shi, P., Li, Q., & Chen, Z. (2018). A study on influence of magnetic field on fracture of ferromagnetic material due to magneto-mechanical coupling effects. International Journal of Applied Electromagnetics and Mechanics, 1–11. doi:10.3233/jae-171156
12. Xu, J., Li, Y., & Chen, G. (2019). Effect of Tensile Force on Magnetostrictive Sensors for Generating and Receiving Longitudinal Mode Guided Waves in Steel Wires. Journal of Sensors, 2019, 1–8. doi:10.1155/2019/9512190

13. J.L. Ren, K. Song, H. Zhu, N. Tang, Application of magnetic memory testing in ferromagnetic items of power station, in: *Proceedings of the SICE 2003 Annual Conference, Fukui, Japan,* 4–6 August, (2003) DOI: 10.1109/sice.2003.1323642
14. Z.D. Wang et al. / Journal of Magnetism and Magnetic Materials 324 (2012) 382–388
15. Bulte, D. ., & Langman, R. . (2002). Origins of the magnetomechanical effect. Journal of Magnetism and Magnetic Materials, 251(2), 229–243. doi:10.1016/s0304-8853(02)00588-7
16. Zhang, X., Zhao, M., Huang, Y., & Zheng, Y. (2018). Novel non-contact torque measurement using the magnetomechanical effect. Instrumentation Science & Technology, 1–10. doi:10.1080/10739149.2018.1474767
17. Zhang, J., Wang, D., Song, D., Zhang, D., Zhang, C., Wang, D., & Araújo, J. A. (2019). Tribo-fatigue behaviors of steel wire rope under bending fatigue with the variable tension. Wear. doi:10.1016/j.wear.2019.03.004
18. Ferreira, N. J., Antunes, P. J., Viana, J. C., Fernandes, L. C., Fernandez, E., & Lanceros-Mendez, S. (2019). Magnetic Field Perturbations by Thermoelectric Effects. 2019 ESA Workshop on Aerospace EMC (Aerospace EMC). doi:10.23919/aeroemc.2019.8788964
19. SpinMeter-3D USB 3 Axis Magnetometer – the maintenance instruction

Influence of rare earths on biodegradable magnesium alloys used in medical prostheses

Adriana SAVIN, Rozina STEIGMANN, Gabriel Silviu DOBRESCU

National Institute of R&D for Technical Physics, NDT Department, Iasi, Romania

Abstract: Mg alloys have been designed to meet the requirements of bone repair implant materials by adding Calcium and Yttrium. Usually Ca is added to control corrosion rate of Mg alloys and thinning grain boundaries. The system is defined as Mg-0.5Ca-xY (x=0.5, 1, 1.5), varying the Y concentration in order to slow the degradation process. This morphological characterization with SEM, EDX, noninvasive testing is required to be carried out the determination of mechanical characteristics. The paper present the influence of Yttrium over elastic properties of these alloys in order to choose the best values appropriate with human bones, using Resonant Ultrasound Spectroscopy and ultrasound method.

Keywords: biodegradable materials, magnesium alloys, rare earths, ultrasonic methods

1. Introduction

The importance of biomaterials has increased once with the innovations in medicine, especially those in surgical techniques that have made possible to use biomaterials for replacing body parts that have lost their function due to illness, aiming at improving the function and correcting the abnormality. As biodegradable material [1], the implant has the properties of degrading in the same time with the surrounding tissues or after the accomplishment of the role for which it has been used.

Magnesium as degradable material [2] has been used first time for the ligaments of blood vessels, later being used as material for osteosynthesis [3]. Calcium (Ca) and zirconium (Zr) have a positive effect on mechanical properties at low concentrations of 2wt% and respectively 0.42wt%, while rare earth (RE) can form intermetallic phases with Mg, improving the mechanical strength of the alloy. Mg alloys used in medical applications have the selection of alloying elements not only based on their ability to improve mechanical properties and corrosion rate, but also on their biocompatibility degree. Once released into the body by the alloy degradation process, the alloying elements must be completely reabsorbed without producing local or systemic toxic effects. From all of these, Al is not compatible with use in medical applications due to toxic effects. Zn, Mn, Li and RE can induce toxic effects at certain concentrations. Ca it shows the best biocompatibility, no toxic effects, being, as Mg, an element present naturally in the human body. It is known that pure Mg has a poor mechanical performance, the mechanical requirements of the implant daily cannot be satisfied [2,4,5], and in addition Mg alloys have low corrosion resistance and therefore degrade rapidly in the human body. Thus, the development of new types of Mg alloys using new alloying elements has emerged as a natural necessity.

The Mg Ca Y alloy has been studied and tested for biodegradable materials. The impact of Y on mechanical properties followed by biological analysis is an important step in the study, with major concerns over the use of Mg alloys for carrier implant materials.

This paper proposes to analyses the influence of yttrium concentration over elastic properties using Resonant Ultrasound Spectroscopy (RUS) and ultrasound method (as noninvasive methods) to examine mechanical properties (i. e. Poisson ratio, Young modulus, Shear modulus) of biomaterials based on magnesium alloy Mg 0.5Ca xY with different ratios composition of Y(x= 0.5; 1.0; 1.5; wt%;), RUS assuring an evaluation on the entire material mass, and by EDX and SEM, the chemical structure of the studied alloys has been established.

2. Materials and Methods

The Mg 0.5Ca xY having selected with high purity elements as raw materials (Mg-99.7%,Mg-15Ca and Mg-30Y master alloys) were melted in an induction furnace under argon atmosphere in a graphite crucible [6, 7]. The final chemical composition 0.5Ca and xY (determined as an average of at least four values for same sample was performed by repeated casting operations and was analyzed by EDX using SEM Quanta 200 3D DUAL BEAM coupled with an EDS-EDAX detector.

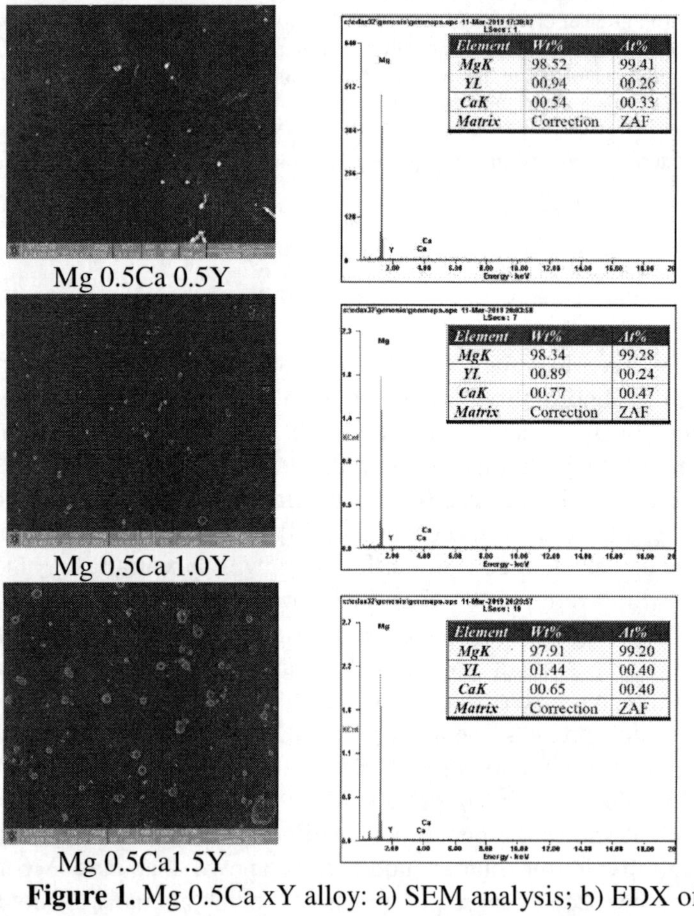

Figure 1. Mg 0.5Ca xY alloy: a) SEM analysis; b) EDX of specimen

A further EDX analysis of chemical composition of Mg0.5Ca sample (Figure 1b) confirms that no other elements except Mg, Ca and Y are present in the alloy. SEM images (Figure 1a) emphasize that the microstructure of the Mg0.5xY alloys are semi-compacted, may be yttrium grains remain still remain unfused. The cube samples, with 10 mm side, has been cropped using MAXIEM 1530 have been studied and probes was embedded in resin prepared using EpoFix Resin Struers product. The samples surfaces have been prepared for metallograpy analysis using DiaPro 3÷1μm diamond suspensions and Tegramin 30 equipment for grinding and polishing. Hardness was measured with NOVA 360 instrument for indentation time 10seconds and $5x10^3$gf. Microscopic analysis was obtained used AxioCam MRc5-Zeiss.
For calculated mechanical parameter was used ultrasound method by echo pulse method, described in [8] in order to determined Young modulus E, shear modulus G and Poisson ratio of sample using longitudinal velocity wave C_l and transversal velocity wave C_t presented in table 1. The average values of hardness were obtained from five different areas of faces of

sample remains approximatively constant. The small differences are due to the influence of alloying elements that influence compactness.

Table 1. Sample characteristics

Sample	Y Comp. (%wt)	length [mm]	Density [g/mm^3]	Young modulus [GPa]	Shear modulus [GPa]	Poisson ratio	C_l [m/s]	C_t [m/s]	HV5
#1	0.5	9.94	1649.84	112.96	37.79	0.28	5282	2948	52.8
#2	1.0	10.04	1693.77	118.44	36.50	0.31	5442	2868	38.9
#3	1.5	10.19	1696.78	88.81	36.16	0.31	5610	2931	37.5

*the sample present a crack at surface

Resonant ultrasound spectroscopy (RUS) as method has the ability to emphasize the modification of resonance frequencies for a given material. The natural frequency depends on the elasticity, size and shape of the object - RUS exploits this solid property to determine the elastic tensor of the material. These frequencies depend on the size, shape, density and elastic properties of the sample. Each frequency corresponds to a particular fascicle of propagating waves and the exchange of modes [9]. These maxims should be repeated at 1/f intervals, where f is the resonance frequency. A sufficient set of experimental data on the measured resonance frequencies can provide information about the sample properties that have been tested.

3. Experimental set-up, results and discussions

Methods based on the time of flight measurement evaluate the elastic wave velocities that occur in the material as well as their attenuation. Making these determinations involves using waves with a wavelength much smaller than sample sizes. For these samples, a signal delay block was used. The longitudinal velocity wave C_l and the transverse velocity wave C_t presented in Table 1 were determined by the experimental assembly, Figure 2a.

a b

Figure 2. Experimental set-up: a). Ultrasound determination of propagation velocities; b) RUS measurement

RUS involves exploring the resonant structure of a compact sample such as cube, parallelepiped, or short cylinder. In the RUS method, ultrasounds were generated and received using two compressing waves transducers P111.O.06P3.1 US type with 60kHz central frequency. A transducer produces vibration and the other converts the resonance motion into an electrical signal. The sample is supported by the contact force; it does not require gluing or coupling agents.

The frequency sweep performed was between 80kHz and 320kHz in 100Hz steps using a Network /Spectrum/Impedance Analyzer (NSIA) type 4395A Agilent USA. The signal generated by NSIA is amplified using a broadband Amplifier A1012AG TC Power Conversion Inc. and is applied to an ultrasound transducer. The signal provided is received by an identical transducer and applied to the analyzer port B, to be further processed by the PC.
Most of the natural frequencies of an oscillating cube do not respond to analytical solutions. By performing numerical simulations (for samples of similar dimensions to those taken in the study, in a wide variation range of G and ν) it resulted that always the first pic corresponds to a torsional mode which, in addition, is the fundamental torsional mode. For such a mode, the fundamental frequency of the mode is given by $f = \frac{m}{2d}\sqrt{\frac{G}{\rho}}$ where f is the measured frequency, d is the height of the sample, G is the shear modulus, ρ is the density and m is an integer. For fundamental mode, m=1.

Under these conditions, G = 36.16GPa for sample # 3. Using procedure indicated in [9] and taking G at the determined value, the value of the Poisson coefficient ν was swept within a reasonable limit in accordance with the values determined between 0.28 and 0.31, which implies the modification of the C_{11} and C_{12} elements in the elastic matrix C. For each the value of ν, the frequencies of the ultrasonic resonance spectrum modes have been determined by numerical simulations, selecting that value of the Poisson coefficient that minimizes the function $F = \sum_{i=1}^{N} w_i \left(f_i^c - f_i^m \right)^2$, where N is the number of peaks in the spectrum, f_i^c is the ith calculated frequency; f_i^m is the measured and w_i is the weighting factor. The oscillation eigenfrequencies were calculated based on the parameters set presented in Table 1 using SolidWorks 2014, Simulations Toolbox.

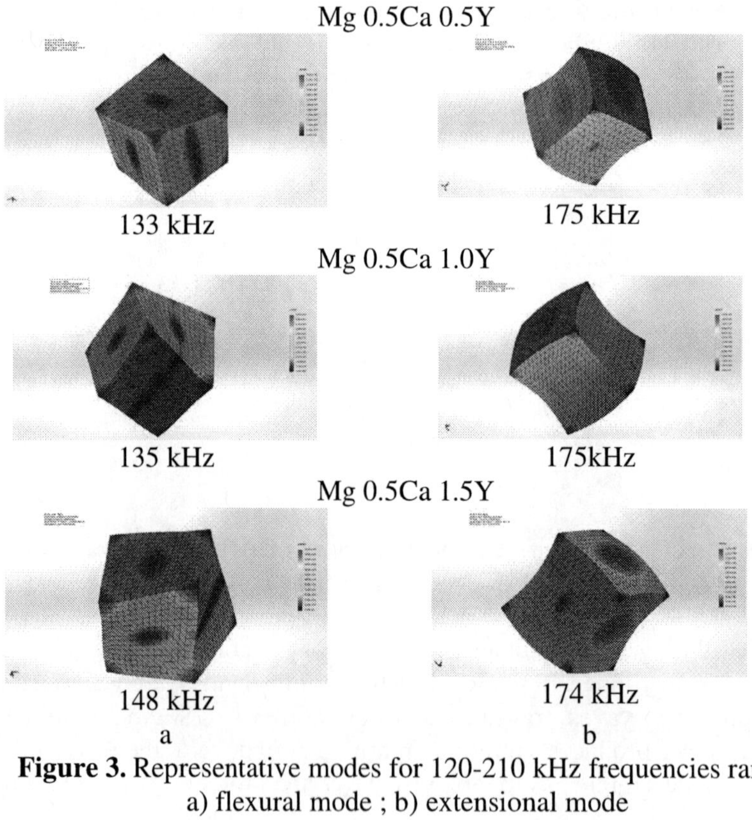

Figure 3. Representative modes for 120-210 kHz frequencies range:
a) flexural mode ; b) extensional mode

These simulations are required to determine the frequency range sweep. The RUS spectra for samples # 1- # 3 in the frequency range 120kHz-220kHz were plotted (Figure 4). The samples are fixed between transmitter and receiver transducers in corners so that the condition of the surface is stress-free. At the same time, the grip on the corners ensures the condition of obtaining a large number of resonant frequencies.

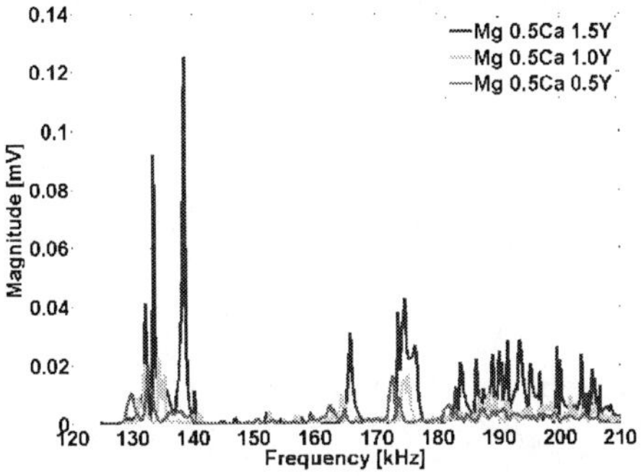

Figure 4. Resonance ultrasound spectra for the sample

The inhomogeneities from sample can be identified from the resonance spectrum, due to the resonance frequencies swift, peaks splitting, increasing of height and magnitude modification. The method is based on the estimation of resonant eigenfrequencies [10], based on an eigenvalue and eigenfunction method [11].

It can be noticed that by the presence of anomalies within the samples (pores or pores clusters) the RUS spectrum changes in both amplitude and phase. In this case, it is not the question of determining the elastic size by the proposed method but only by a data inversion, considering that the matrix of the model is not well conditioned.

3. Conclusions

Magnesium alloy used as biodegradable materials display a variety of properties and depends on the chemical composition and biocompatible elements that are included. Addition of calcium and yttrium at magnesium alloys improve the mechanical properties, yttrium contributed to obtained special corrosion resistance.

The Mg-based alloys were obtained in an induction arc melting furnace in argon atmosphere, and were characterized (microstructural and chemical analysis) with SEM Quanta 200 3D with EDX. Morphological analysis using EDX as well as microidentation was required to correlate the data obtained with those proposed by the research, i.e. the proximity of mechanical parameters of the alloy to the human bone.

The ultrasonic resonance allowed the determination of the Young modulus, the shear modulus and Poisson ratio of the sample, starting from data characterizing the material in the entire volume, not only at the surface, using longitudinal velocity wave and transverse velocity waves.

Acknowledgements

This work was supported by the Romanian Ministry of Research and Innovation, CCCDI – UEFISCDI project PN-III-P1-1.2-PCCDI-2017-0239, Nucleus Program PN 19 26 01 02 and Project 11 PFE/2018.

References

[1] Prakasam M, Locs J, Salma-Ancane K, Loca D, Largeteau A, Berzina-Cimdina L 2017 *J. Funct. Biomater.* **8**(4), 44

[2] Witte F, 2010 *Acta biomaterialia* **6**(5) 1680-92

[3] Chorny V, Yatsun Y, Golovakha M 2014 *Orthopaedics, Traumatology and Prosthetics* **1** 105-9

[4] Staiger MP, Pietak AM, Huadmai J, Dias G. 2006 Biomaterials **27**(9) 1728-34

[5] López HY, Cortés-Hernández DA, Escobedo S, Mantovani D 2006 *Key Engineering Materials* **309** 453-6

[6] Lupescu S, Istrate B, Munteanu C, Minciuna, M G, Focsaneanu S, Earar K 2017 *Revista de Chimie* **68**(6) 1408-1413

[7] https://erris.gov.ro/Centrul-de-Cercetare-in-Ingi-2, last seen on April 19, 2019

[8] Savin A, Craus ML, Turchenko V, Bruma A, Dubos PA, Malo S, Konstantinova T, Burkhovetsky V 2015 *Applied Sciences* **5**(4) 1665-82

[9] Savin A, Vizureanu P, Prevorovsky Z, Chlada M, Krofta J, Baltatu MS, Istrate B, Steigmann R. 2018 *IOP Conf Ser: Mater Sci Eng* **374**(1) 012030

[10] Myasnikov DV, Konyashkin AV, Ryabushkin OA 2010 *Tech. Phys. Lett.* **36**(7) 632-5.

[11] Jaglinski T, Lakes RS 2011 *Rev. of Sci. Instrum.* **82**(3):035105

Arcan device employed in CFRP testing

Andrei-Narcis DANILA[1,2], Rozina STEIGMANN[1],
Adriana SAVIN[1], Igor BLANARI[1,2], Paul Doru BARSANESCU[2]

[1] [1] Nondestructive Testing Department, National Institute of R&D for Technical Physics, Iasi, Romania
[2] Faculty of Mechanical Engineering, University of Ljubljana, Ljubljana, Slovenia
[3] Faculty of Mechanical Engineering' Technical University Gh.Asachi, Iasi, Romania

Abstract: The study investigates the behavior at mechanical loading tests of laminate composites manufactured by two methods as wet lay-up with vacuum bagging and autoclave processing method and they were achieved using two types of epoxy/carbon laminates. The stacking sequence of layers is imposed at eight-ply laminate $[0]_8$ and $[(45/0)_2]_s$. Fibers orientation, stacking sequences and methods of production influence the response of the composite structure. Starting from estimating the response of the composite structure to mechanical loads, an optimized Arcan device for testing thin composite plates was developed and used. Arcan type tests are used to investigate the specimen's behavior in pure normal, pure-shear and mixed modes respectively. The fixture allows an ARCAN sample to be rotated at 15° increments from 0° to 90° to produce different amounts of mode-mix during loading; with 0° being pure tension/compression loading to 90° being pure shear loading. The optimization of the design of the proposed device, based on simulations with finite elements and the experimental results obtained, completes the analysis regarding the performances of the composite materials processed with autoclaves, justifying high investment for improvement of material properties.

Keywords: Arcan device, CFRP, mechanical tests, FEM, complex stresses

1. Introduction

The specific degradation of CFRP consists of delamination due to impacts, even at low energies, or overloading of the structure; excessive porosity; water absorption [1] etc. Evaluation of impact damage mechanism of epoxy composite structures is approached in numerous studies [2, 3]. Because assemblies of composite modules and connections between composite and metallic structures are of great importance, few standard tests are necessary to characterize composite assemblies [4]. Failure initiation in such assemblies is associated with the apparition of cracks in adhesive or interlaminar zones close to joints of composite-to-composite or composite-to-metal, due to the efforts caused by efforts supported differently by the components of the joint [5]. The specialized literature proposes numerous experimental devices and often a large number of specialized specimens for each type of composite application, in accordance with the applicable domain standards [6].
The use of an Arcan device to determine the behavior of composites for hybrid assemblies and beyond, has the advantage of testing of thin composite plate, cross ply and quasi-isotropic symmetrical plain weave carbon-epoxy laminates, for a particular set of epoxy / carbon fabric type. The optimization of the design of the test device and of the specimen was carried out based on the simulations of FEM using ANSYS Mechanical academic version [7]. From this reason, the failure of structural components from CFRP, submitted to complex axial stresses is more studied nowadays. The test (V notched beam) is used on a wider scale for pure shear, and it is also standardized to be applied to different materials [8]. Axial loading for CFRP shaped sample is realized using an Arcan device [9]. The CFRP samples taken in the study was manufactured by two methods, wet lay-up with vacuum bagging and autoclave processing method and they were achieved using two types of epoxy/carbon laminates [10].
The aim of this paper is to present the complementary results obtained during the investigation of mixed-mode fracture using the Arcan device for tests based on the sample.

2. Materials and methods

For these tests, two type of carbon/epoxy laminates, manufactured by wet lay-up and vacuum bagging and autoclave processing method with different ply stacking sequences $[0]_8$ and $[(45/0)_2]_s$ [10] were realized. Dimensions of plates were 210x297mm with 2-2.8 mm thickness. For tension and compression tests, the results were those from [10].

The complementary method proposed for the complete characterization of the obtained materials is the use of a modified Arcan test. Arcan type tests are an easy way to study the specimens behavior in pure and mixed mode (tensile & shear) respectively. The fixture allows an ARCAN sample to be rotated at 15° increments from 0° to 90° to produce different amounts of mode-mixt during loading; with 0° being pure tension/compression loading to 90° being pure shear loading. The key advantage of this device is the testing of composite thin plates. The characterization of the material from elastic properties point of view follows

$$E_1 = E_f V_f + E_m V_m \tag{1}$$

with E_f the elastic modulus of the fibers; E_m the elastic modulus of the matrix; V_f the fiber content; V_m the volume content of matrices.

The strain - stress responses of laminate $[0]_8$ and $[(45/0)_2]_s$ in tension, was determined [10] and are presented in Table 1, where E is elastic modulus; X_1^T tensile strength (σ is stress); ε is strain. For composite structures, tensile failure of a 0° ply is usually represented by a maximum – strain failure initiation criterion $\sigma_1 = X_1^T$ or $\varepsilon_1 = e_1^T$.

Table 1: Results of static tests in tension and compression

		Method Stacking Sequence			
		Wet lay-up + vacuum bagging		**AC-prepreg**	
		$[0]_8$	$[(45/0)_2]_s$	$[0]_8$	$[(45/0)_2]_s$
d (mm)		2.80	2.78	2.05	2.04
V_f (%)		32	31	45	44
Tension	E_1[GPa]	40.9		56.4	
	X_1^T[MPa]	398		527	
	ε_1^T[%]			0.945	
	e_1^T[%]	0.95			
	E_x[GPa]		27.8		46.9
	σ_x^T[MPa]				404
	ε_x^T[%]				1.073
	σ_x^{ult}[MPa]		292		
	ε_x^{ult}[%]		1.125		

It can be observed that the elastic modulus and strengths are smaller in wet lay-up laminates for all loading cases. The autoclave processed Torayca® T300 fibres /ET445 prepreg samples exhibited significantly higher tensile, compressive and flexural strengths (24-42%) than lay-up samples [10]. The laminates were fabricated using standard modulus carbon fiber plain weave fabrics, Sigrafil and Torayca. Both fibers in fabrics are polyacrylonitrile-based, with equal filament diameter of 7 μm, the number of filaments in yarn being 50k and respective 3K filaments) and nearly similar elastic modulus [11],[12]. For the matrix, the MGS epoxy system (trade name MGS® 285, type diglycidyl-ether of bisphenol) and polyamide-type hardener (trade name Epikure Curing Agent MGS LH 285) were used [13].

2.1. Test fixture and specimen

In order to produce different biaxial states of stress, for this study has been used a modified Arcan test fixture [14]. Using this test fixture, various biaxial states of stress can be obtained simply by varying the angle α between longitudinal axis of symmetry of the specimen (mounted in the test device) and the loading direction, as follows tension or compression is obtained for $\alpha=0°$, shear for $\alpha=90°$ and tension (or compression) with shear for other values of the angle α (Figure 1). The Arcan device and method, allow the application of shear stress ($\alpha=90^0$), tensile ($\alpha=0^0$), or shear with traction (with different ratios between σ and τ) in the narrow section by changing the angle of application of force, Figure 1.

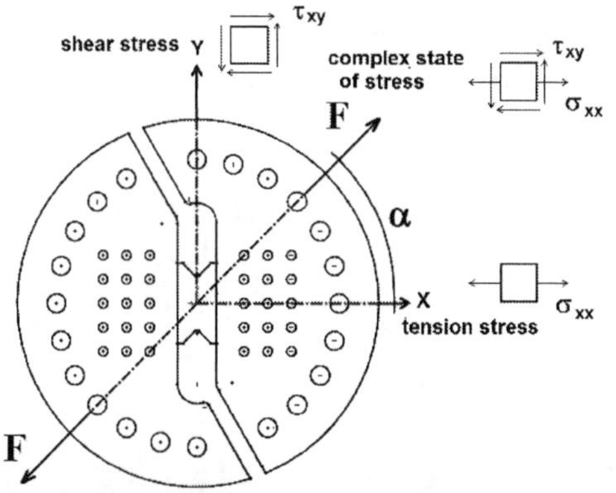

Figure 1. Arcan fixture for shear test with different loading configuration

The shape of the sample concentrates the stress towards its center, where is the smallest cross section. In order to verify this assumption, a FEM analysis was done for one sample mounted on the Arcan device, under different cases of loading.

The two types of specimens (modified butterfly type specimens) are presented overlapped in the Figure 2. Note that the central part (where is the minimum section, in which the rupture occurs) is identical for both specimens. Using FEA and IR thermography [14] in literature is shown that for radius from the bottom of the notch of 2.54 mm the most uniform stress shear stress in specimen is achieved. From the specified plates, samples were prepared [15] in order to study the mechanical behavior of the materials under shear loading (shear strength, shear modulus, strain to failure) fixed to Arcan device. Parallelepiped samples with two notches at the middle of 90^0 each with a depth of 26% of the specimen width are shown in Figure 2.

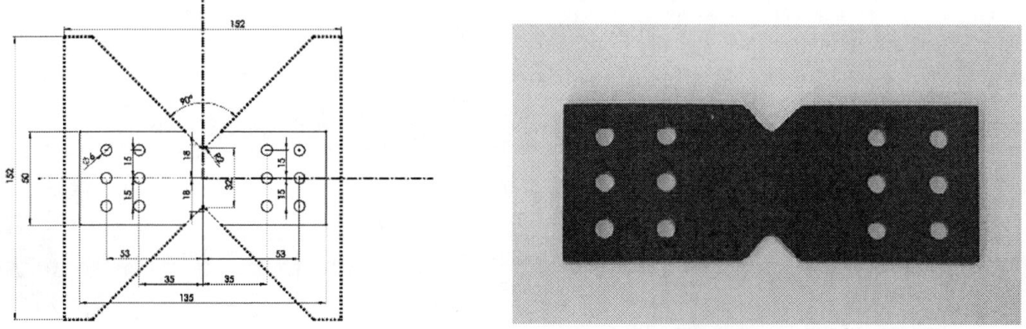

(a) (b)

Figure 2. Specimen: (a) scheme; (b) photo

The geometry of the specimens was selected for positioning in the Arcan device to apply the loads. In order to evaluate the effect of the joint geometry, the distances between the central part and the center of the two rows of parallel holes, realized on both sides of the central area, according to the geometry presented in Figure 2 were determined.

2.2. Finite model analysis

Finite element analyses were performed to determine the state of stress in narrow section of the specimen subjected to shear loads. Using FEM models, the influence of the radius of curvature of the specimen (considered orthotropic) on the uniformity of the stresses that appear in the central section was determined [14]. The sample is fixed at half geometry at Ox direction for prevented body motion. The geometry model transferred in the FEA software and meshed quadrilateral 8 nodes mesh elements as shown in Figure 3. The XY symmetry plane was constrained in the OZ direction. Simulations were performed for sample thickness average of 2.5mm. The FEM mesh contained 287893 nodes and 94984 elements for butterfly specimen (figure 3a) and respective 173593 nodes and 57080 elements (figure 3b) for modified specimen. In the narrow section, the discretization is made similarly for both specimens.

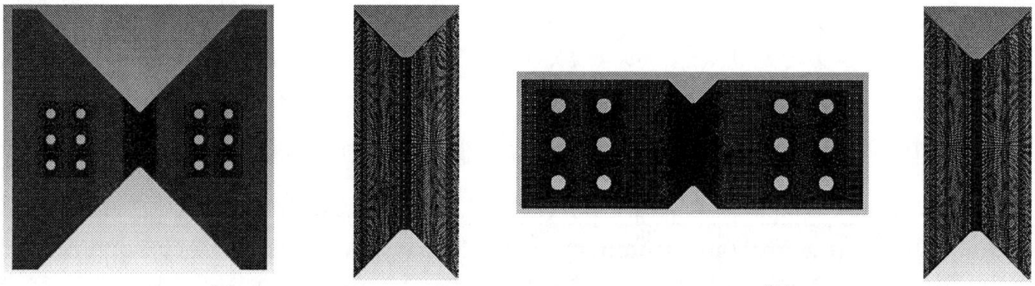

(a) (b)

Figure 3. Mesh applied to studied specimens: (a) butterfly; (b) modified specimens

The results of FEM for the first type of material are presented in Figure 4.

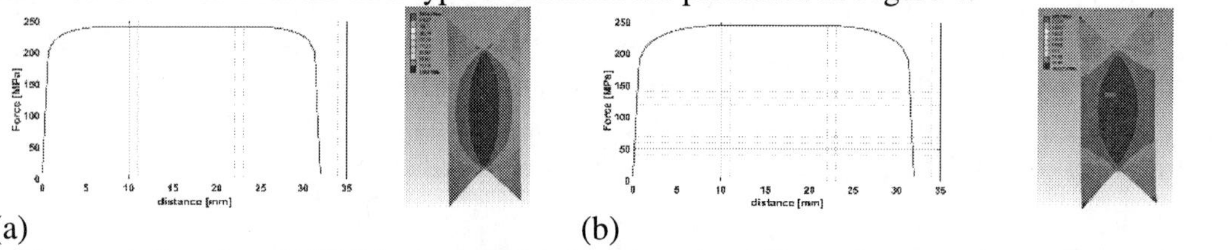

(a) (b)

Figure 4. Results of FEM for material 1 for Linearized Shear Stress(XY plane) and Shear Stress (XY Plane): (a) butterfly specimen;(b) modified specimen

The results of FEM for the second type of material are presented in Figure 5.

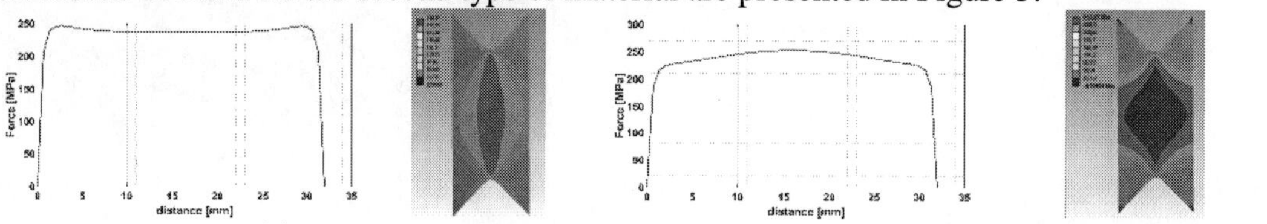

(a) (b)

Figure 5. Results of FEM for material 2 for Linearized Shear Stress(XY plane) and Shear Stress (XY Plane): (a) butterfly specimen;(b) modified specimen

The similar distribution of stresses in both the butterfly and the modified specimens, in the narrow section promotes the modified specimen to be used in the tests specified below.

3. Results and discussions

The Arcan device was mounted on a *WDW 50 testing* machine with a maximum load of 50 kN, Figure 6a. The loading applied to the specimens was carried out at a transversal speed of 2 mm/min. The samples cropped from the four plates have been tested using the following values of the angle α [45°, 60° and 90°]. The experiments were performed in a displacement control mode presented above. Shear strains were extracted from the equipment at the center of each specimen.
Figures 6b and c show two typical failed specimens.

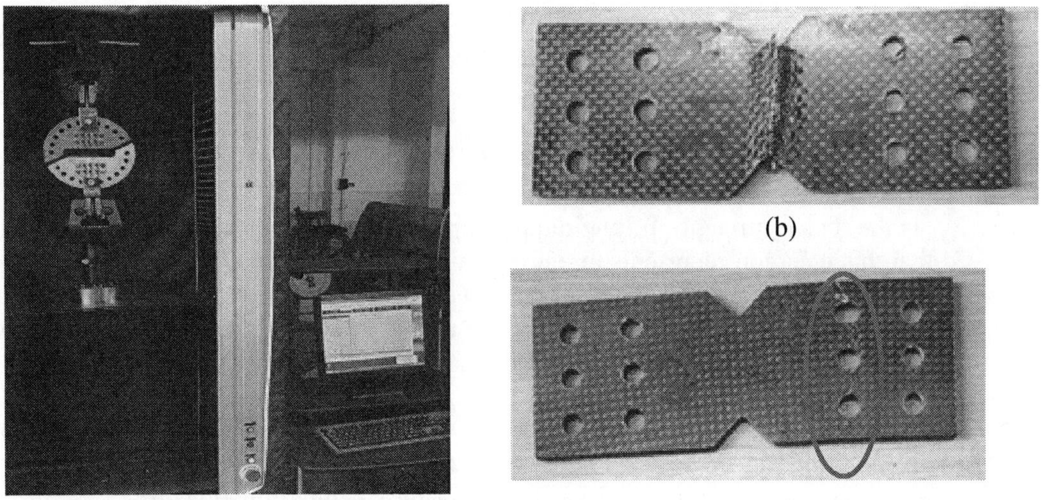

(b)

(a) (c)

Figure 6. Experimental set-up: (a) Arcan fixture; (b) failure of specimen in the narrow section; (b) failure of specimen after reaching the ultimate stress, by cracking between holes (red circle)

The analysis of the specimens after the test showed that the rupture of the samples is influenced by the fiber distribution and the force direction (α angle). The results obtained for the specimens that did not break in the narrow section were not taken into account, this phenomenon being also pointed out in the literature for the butterfly specimens [14]. Shear strengths from ARCAN tests are presented in Table 2.

Table 2: Results of pure shear and mixed shear/tension ARCAN test

Factor of safety c_1 (calculated for ellipses drawn with dotted line)				
Material	**σ [MPa]**	**τ [MPa]**	**X_{11} [MPa]**	**X_{12} [MPa]**
P1	149.03	149.03	527	214.11
P2	115.25	115.25	404	173.13
P3	92.95	92.95	398	159.79
P4	101.57	101.57	292	153.75

4. Tsai Hill Failure criterion

Halpin and Tsai [16] have been developed few simple models for calculation of elastic properties for composite materials based on their elasticity. The composites with long fibers are orthotropic. In the absence of specific analytical relationships for these materials, the stress state can be studied using numerical methods. Tsai-Hill criterion is based on an energy theory and it is widely used for orthotropic composite materials. For a general biaxial state of stress (biaxial tension with shear) it is usually written as

$$\left(\frac{\sigma_{11}}{X_{11}}\right)^2 - \left(\frac{\sigma_{11}\sigma_{22}}{X_{11}}\right)^2 + \left(\frac{\sigma_{22}}{X_{22}}\right)^2 + \left(\frac{\tau_{12}}{X_{12}}\right)^2 \leq 1 \tag{2}$$

where X_{11} and X_{22} are the allowable strength in the longitudinal and respectively transverse direction of composite material; X_{12} is in plane shear allowable strength. The Tsai-Hill criterion predicts the failure when the left member reaches 1. For a particular biaxial state of stress with uniaxial tension and shear the equation will possess only the normal stress σ_{11} (σ_{22}=0) and the rupture will occur when

$$\left(\frac{\sigma_{11}}{X_{11}}\right)^2 + \left(\frac{\tau_{12}}{X_{12}}\right)^2 = 1 \tag{3}$$

where X_{11} is the tensile strength in longitudinal direction; X_{12} is in plane shear strength. It is observed that this equation represents an ellipse in the space σ-τ. The semiaxes of the ellipse are X_{11} and X_{12} respectively. The values of X_{11} were taken from [10], where the tests have been carried out on classical specimens for tensile tests – dog-bones. The materials used with Arcan tests are from the same batch as the materials used for tensile tests in [10]. On the contour of the ellipse, the material failures and the safety factor is c=1. Inside the ellipse is the safe area (c>1) and outside is the dangerous area (c<1). In figures 10 the ellipses generated by the Tsai-Hill criterion are presented for the four composite materials. In each diagram two points at which the composite material has failed for the loading directions α=45° and α=60° are also represented. It is observed that these points are located inside the ellipses (in the area considered safe). Of course, if only for certain states of stress the material failure occurs in the area considered safe, this can create serious problems for calculation in engineering.

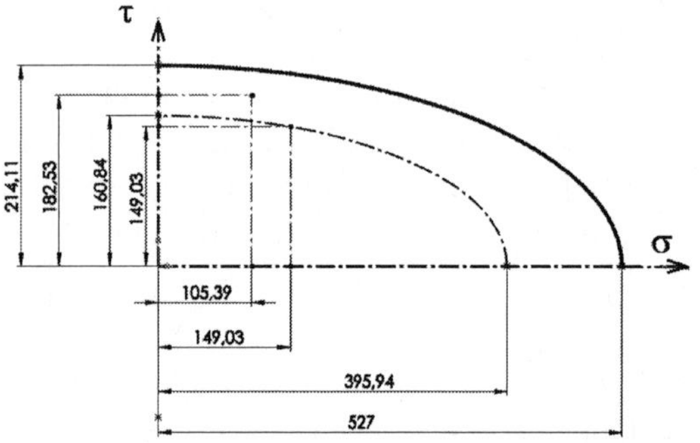

(a)

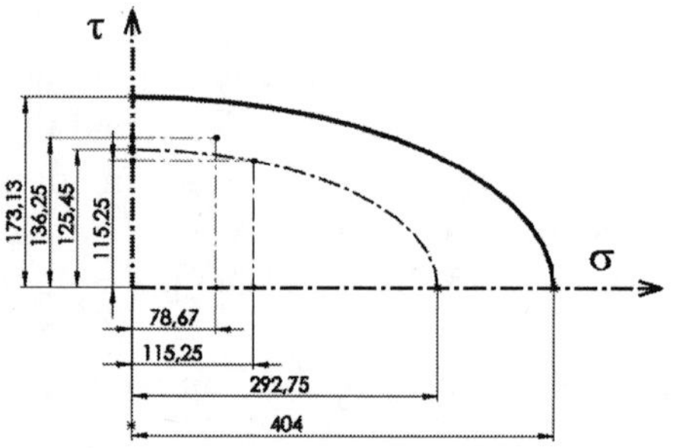

(b)

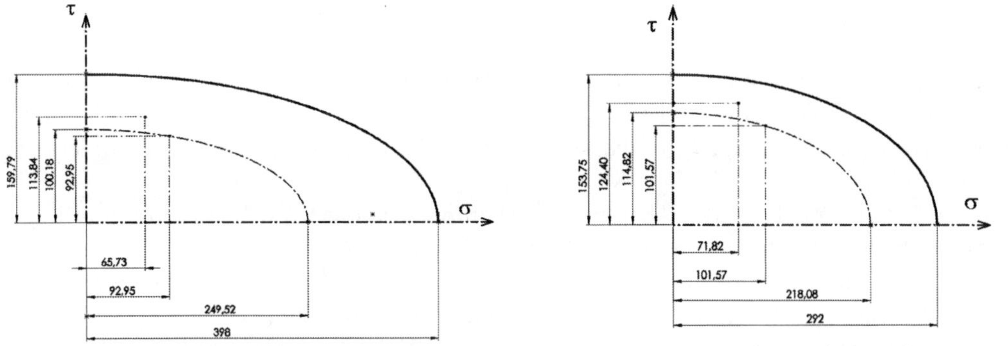

(c) (d)

Figure 7. Tsai-Hill failure criteria for composite structures: (a) plate 1; (b) plate 2; (c) plate 3; (d) plate 4;

This problem can be solved as follows: an ellipse concentric to the first is drawn, so that the states of stress for which the failure of material is produced must be on the contour of the ellipse or outside of it. The semi axes of this ellipse are X_{11}/c_1 and X_{12}/c_1 respectively, where is a factor of safety ($c_1>1$). Thus the safe area will be considered now the one *inside smaller ellipse*. The semi-axes of this ellipse are X_{11}/c_1 and X_{12}/c_1 respectively, where is a factor of safety ($c_1>1$). In Table 3, the factor of safety c_1 are presented for the four composite materials taken into consideration.

Table 3. The factor of safety c_1

Material	c_1	**The semi-axes of the ellipse which delimits the area considered safe [MPa]**	
		X_{11}/c_1	X_{12}/c_1
P1	1.331	395.94	160.86
P2	1.380	292.75	125.45
P3	1.595	249.52	100.18
P4	1.339	218.07	114.82

Thus the safe area will be considered now the one inside smaller ellipse and Tsai-Hill criterion can be easily rewritten for each composite material studied in this work. The obtained results show that the failure criteria taken into consideration are correct and can be used for the prediction of failure in the analysed composite structures.

5. Conclusions

Four different composite materials were tested using a modified Arcan test fixture and modified butterfly specimens, up to ultimate failure. It has been shown that the Tsai-Hill criterion can be used for the studied composites, in a biaxial state of stress. At the failure study, the following factors must be taken into consideration, thickness of the lamina, stacking sequences of the laminates, pretension of fixing elements, the method for obtaining of laminates, which affect the tensile failure in different ways. Thus, the testing at complex stresses states is important to transition of one failure mode to another.

Authors proposes a testing program in mixed configuration for Arcan in order to selective investigate few critical parameters. The tests on failure mechanism during the mixed loadings will be further developed. Supplementary researches of these specimens for failure avoidance around fixing elements are imposed.

References

[1] Sang, L., Wang, Y., Wang, C., Peng, X., Hou, W. and Tong, L., 2019, "Moisture diffusion and damage characteristics of carbon fabric reinforced polyamide 6 laminates under hydrothermal aging," *Composites Part A:* Applied Science and Manufacturing, **123**, pp.242-252.

[2] Saito, H., Kimpara, I., 2006, "Evaluation of impact damage mechanism of multi-axial stitched CFRP laminate," *Composites: Part A*, **37**, pp.2226 – 2235.

[3] Garnier, C., Pastor, M.L., Lorrain, B. and Pantalé, O., 2013, "Fatigue behavior of impacted composite structures," *Composite Structures*, **100**, pp.443-450.

[4] Jiang, X., Qiang, X., Kolstein, H. and Bijlaard, F., 2016, "Analysis on adhesively-bonded joints of FRP-steel composite bridge under combined loading: Arcan test study and numerical modeling," *Polymers*, **8**[1], p.18.

[5] Totry, E., González, C. and LLorca, J., 2008, "Failure locus of fiber-reinforced composites under transverse compression and out-of-plane shear," *Composites Science and Technology*, **68**[3-4], pp.829-839.

[6] Mespoulet, S., Hodgkinson, J.M., Matthews, F.L., Hitchings, D. and Robinson, P., 2000, "Design, development, and implementation of test methods for determination of through thickness properties of laminated composites," *Plastics, rubber and composites*, **29**[9], pp.496-502.

[7] ANSYS 14.5 November 2013, "ANSYS 14.5-Mechanical User's Guide" ANSYS Inc.

[8] Lara-Curzio, E., Bowers, D. and Ferber, M.K., 1996, "The interlaminar tensile and shear behavior of a unidirectional C- C composite," *Journal of nuclear materials*, ***230***[3], pp.226-232.

[9] Arcan, M., Hashin, Z.A. and Voloshin, A., 1978, "A method to produce uniform plane-stress states with applications to fiber-reinforced materials," *Experimental mechanics*, ***18***[4], pp.141-146.

[10] Bergant, Z., Savin, A. and Grum, J., 2018, "Effects of manufacturing technology on static, multi-frequency dynamic mechanical analysis and fracture energy of cross-ply and quasi-isotropic carbon/epoxy laminates," *Polymers and Polymer Composites*, **26**[5-6], pp.358-370.

[11] Technical Information. Sigrafil® continuous carbon fiber tow at https://www.sglcarbon.com/en/markets-solutions/material/sigrafil-continuous-carbon-fiber-tows/

[12] Technical information, CC202 ET445, 43% T300 fibre, plain weave 204 gsm, at https://www.toraycma.com/file_viewer.php?id=4462.

[13] Technical Information for laminating resin MGS L285 at https://www.hexion.com/en-us/chemistry/epoxy-resins-curing-agents-modifiers/epoxy-tds

[14] El-Hajjar, R. and Haj-Ali, R., 2004. In-plane shear testing of thick-section pultruded FRP composites using a modified Arcan fixture. Composites Part B: Engineering, 35[5], pp.421-428.

[15] ASTM D5379 / D5379M-12 2012 Standard Test Method for Shear Properties of Composite Materials by the V-Notched Beam Method, ASTM International, West Conshohocken, PA.

[16] Halpin, J.C., 1969. Effects of environmental factors on composite materials (No. AFML-TR-67-423]. Air Force Materials Lab Wright-Patterson AFB OH.

Experimental setup for MPI sensitivity measurements: First results

Pavel STANĚK [1], Zbyněk ŠKVOR [1]

[1] Department of Electromagnetic Field, Faculty of Electrical Engineering, Czech Technical University in Prague, Technická 2, 166 27 Prague 6, Czech Republic
Phone: +420 22435 2278, e-mail: stanepa2@fel.cvut.cz, email: skvor@fel.cvut.cz

Abstract
During the development process of MPI equipment it is very helpful to know the detection ability of the MPI process. Such an ability depends on material properties of the body under the test, applied magnetic field, defect shape and size, and detection particle properties. There is not enough information published, therefore in-house experiments are justified.
This paper aims to present an experimental setup for MPI sensitivity measurement. The setup consists of four coils and two steel plates attached to the coils. The steel plates are magnetized by the flux generated by the coils. The coils are situated at the vertices of a square and the opposite coils are electrically connected in series therefore the setup can be fed by two independent current waveforms. These current waveforms generate two perpendicular magnetic field components on the plates so it is possible to generate arbitrary tangential magnetization on the surfaces of the steel plates. A standard shim gauge is placed on the steel plate in the middle of the coils. If the setup is fed by two sine waveforms of the same frequency and arbitrary phase shift, the tangential magnetization of the steel plates is elliptical polarized. The coils are fed from a pair of arbitrary-waveform generators amplified by a pair of power amplifiers.
Using this setup the sensitivity of MPI can be measured for various input parameters. Measurement setup as well as first experimental results are presented in this paper.
Keywords: Magnetic particle inspection (MPI), sensitivity, magnetic field polarization

1. Introduction

The performance of a non-destructive testing (NDT) technique can be described by two parameters: sensitivity and reliability. Sensitivity describes the smallest detectable defect by an application of the NDT method. Reliability is defined as a probability that a system will perform its function. Probability of detection (POD) as defined in ref. [1] or in ref. [2] is a statistically evaluated function of probability of a defect detection usually determined as a function of the defect length. Details of application of MPI such as magnetization method, magnetic field direction and strength, and detection particles used are not mentioned in these studies. Only an average performance of MPI is evaluated [1]. The average performance can be also affected by skills or emotional mood of the operators. However, the proper application of all MPI steps is essential for successful detection. The knowledge of sensitivity and reliability is especially useful for developing and optimizing NDT equipment to set optimal conditions and increase the POD to get closer to the ideal case described in ref. [3]. The presented setup can be used for a measurement of sensitivity for different configurations of a MPI process.

2. Setup description

The experimental setup which is in Figure 1. consists of four coils and two circular steel plates attached to the coils. The steel plates are magnetized by the flux generated by the coils situated at the vertices of a square. All four coils are identical with 90 turns each. The magnetic flux is guided to the surface of the plates by steel coil cores. In this four-coil

configuration the opposite coils are electrically connected in series enabling the setup to be fed by two independent current waveforms. The current waveforms generate two perpendicular magnetic field components in the tangential plane on the surface of the plates. It is possible to generate an arbitrary tangential magnetization on the surfaces of the steel plates. A standard shim gauge (3C2-234) [4] is placed on the steel plate in the middle of the coils. If

the setup is fed by two sine currents of the same frequency and arbitrary phase shift the tangential magnetization of the steel plates is elliptical polarized. The coils are fed by a pair of arbitrary-waveform generators amplified by a pair of power amplifiers as shown in Figure 2. The setup can be easily modified into a three-coil configuration. Three-coil configuration can be supplied from three-phase power grid. This configuration generates rotating magnetic field. The steel plates can be easily removed and replaced by plates made of different magnetic materials or plates with different defect(s).

4. Measurement results

For the sensitivity measurement the arbitrary waveform generator was set to generate two 50 Hz sine waveform with 90° mutual phase shift. The magnitudes were increased until the first indication appeared. The plates were magnetized for 15 s by currents of approximately 8.5 A (RMS). Magnetic field components were measured using Hall probes on the surface of

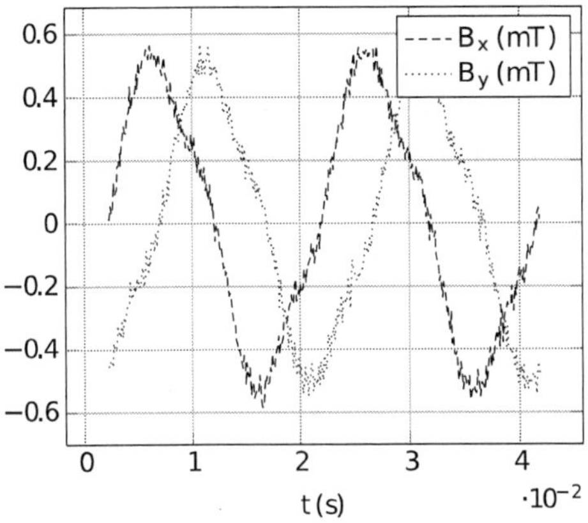

shim gauge placed exactly in the middle. Both magnetic field components are depicted in Figure 3. The corresponding polarization is in Figure 4. Peak-to-peak value of magnetic field components is slightly lower than 1.2 mT which corresponds to 955 kA/m in the air. The magnetic field components were non-sinusoidal despite the sine voltage excitation waveforms at the input of the power amplifier. Higher harmonics were generated due to the presence of magnetic non-linear material (eg. steel). This effect is also responsible for non-circular polarization.

The indication observed on the shim gauge is in Figure 5. There are three circular defects on this type of gauge. The notched defects are of the same width and different depth. The width is 180 μm and the depths are 10 μm, 15 μm, and 20 μm. The circular defect of the largest depth (20 μm) is clearly visible. The indication above the 15 μm depth defect is very weak and is barely visible. Such an indication will be probably missed in a practice. It can be clearly seen that the sensitivity is almost the same in all directions, because the polarization is to the circular polarization. The sensitivity of MPI process for detecting the standard defects on shim gauge starts approximately at 955 $kA/m_{pk\text{-}pk}$ if the plates are magnetized by slightly non-sinusoidal magnetic field components. To evaluate the detection ability more precisely the force impulse on the detection particle should be calculated [6].

The value of measured magnetic field is much lower than the minimal value of the tangential field which should be sufficient to achieve a successful detection according to standards. ASTM standards [4] recommend minimal value of tangential field to be 2.4 kA/m. Slightly lower value 2 kA/m is recommended by ISO standards [5]. The used fluorescent ink was *Chemetall Lumor J*.

4. Conclusion

The setup for MPI sensitivity measurement was built and tested. The sensitivity threshold for a standard shim gauge 3C2-234 was successfully measured. The indications were obtained by magnetic field much weaker than magnetic field recommended by standards.

Acknowledgement
This work was supported by CTU project SGS17/182/OHK3/3T/13.

References

1. S Burke and R Ditchburn, 'Review of literature on probability of detection for magnetic particle nondestructive testing Tech. Rep.', Defence Science and Technology Organisation, 2013.

2. A Zolfaghari and A Zolfaghari and F Kolahan, 'Reliability and sensitivity of magnetic particle nondestructive testing in detecting the surface cracks of welded components', Nondestr. Test. Eval., Vol 33, No 3, pp 290-300, 2018.

3. M Omar, 'Nondestructive Testing Methods and New Applications', IntechOpen, Rijeka, 2012.

4. ASTM International, 'ASTM E709-15 Standard Guide for Magnetic Particle Testing1', 2015.

5. Úřad pro technickou normalizaci, metrologii a státní zkušebnictví, 'ČSN EN ISO 9934-1', 2016.

6. Staněk, Pavel, and Zbyněk Škvor. 'Automated Magnetic Field Evaluation for Magnetic Particle Inspection by Impulse. ' Journal of Nondestructive Evaluation, Vol 38, No 3, 2019.

An initial investigation on the potential applicability of ultrasonic testing to assess effects of corrosion on linear and nonlinear elastic behaviour of pre-stressing steel strands

Karel POSPÍŠIL[1], Marta KOŘENSKÁ[1], Josef STRYK[1], Aleš FRÝBORT[1], Monika MANYCHOVÁ[2]

[1]*CDV -Transport Research Centre, Líšeňská 33a, 636 00 Brno, Czech Republic*
[2]*Brno University of Technology, Faculty of Civil Engineering, Veveří 331/95, 602 00 Brno, Czech Republic*
josef.stryk@cdv.cz

Abstract
Corrosion of pre-stressing strands causes a deterioration of condition of civil engineering structures like bridges, therefore periodic inspections are necessary to prevent collapse of such structures. The corrosion of steel is a slow process however it can be well reproduced in the lab using accelerated corrosion process. Monitoring the amount of corrosion in a sample is of great importance to predict the integrity and the residual life-time of structures. The ultrasonic techniques were proposed as valid methods of analysing the data recorded during the monitoring. In this paper we show results of our experimental study and the application potential of ultrasonic techniques to analyse the effect of damage induced by accelerated corrosion on the linear and nonlinear elastic behaviour of a steel strands. These initial results show the possibility for the application of nonlinear ultrasonic testing methods for monitoring of condition of strands which are built into bridge girders.

Keywords: Electrochemical corrosion, cracking, ultrasonic testing, linear and nonlinear behaviour, pre-stressing steel strands

1. Introduction

Many precast pre-stressed concrete beam bridges were built in the last century throughout world. Pre-stressed concrete with steel strands or rods is a building material closely associated with the advances achieved by building technology and is a widely used method of civil engineering structures. Although steel is usually protected against corrosion in concrete, prestressing steel can suffer stress corrosion.

The life span of pre-stressed concrete infrastructures depends on the corrosion resistance of pre-stressing tendons. The cementations material constitutes a barrier against the ingress of aggressive substances to the steel surface. The high-alkaline pH of concrete pore solution around 12 keeps the steel in passive condition with an oxide passive film formed on the steel surface [1].

However, the diffusion of corrosive species such as chloride and sulphate from external environment through the concrete pores or voided grout in the sheath destroys passivity condition of pre-stressing steel and promotes its corrosion [2]. Another phenomenon, called concrete carbonation, induces the breakdown of the passive layer by reducing the pH value of alkaline pores solution [3, 4].

The stress corrosion cracking (SCC) is the most dangerous strands corrosion form because it can cause the sudden and unexpected failure of steel cables without any externally visible indications [5]. In the past various failures of building structures attributed to SCC. In the present one of the best known is the collapse of bridge structure in Italy. Pre-stressing tendons are buried deeply in the concrete or put in plastic sheath hence corrosion tests based on electrochemical techniques used for reinforced bars are not effective. Acoustic tests are often used, among others, they could be based on either ultrasonic measurements [6] or acoustic emission approaches [7], sometimes applied in combination with other methods [8].

This paper is a continuation of our previous work on detection and characterization of corrosion on concrete reinforcements [9, 10]. The goal of this contribution is to analyze the effects of damage induced by corrosion on the linear and nonlinear elastic properties of a pre-stressing steel strand during the accelerated corrosion process.

2. Tested object and accelerated corrosion procedure

Commercially available seven-wire steel strand of 15.7 mm in diameter was subjected to accelerated corrosion process. The strand consists of a steel central straight wire surrounded by six helical wires (Fig. 1).

Figure 1. Photo of tested seven - wires strand.

The electrolyte used to perform steel corrosion was obtained by dissolving 30 g NaCl in 1 litre of distilled water at room temperature. The corrosion cell was located in the centre of the strand and consisted of a cylindrical plexiglass tube of 60 mm diameter and 16 mm length, fitted with two caps at each side for exits for strand. The corrosion process was accelerated by the impressed current. A potentiostat/galvanostat was used for electrochemical experiments. In such tests, the anode (pole+) is the strand to be corroded and the cathode (pole-) an external electrode [11], in our case made of copper. The steel strand has been polarized under a constant current density $j = 3\ mA/cm^2$. This value has been used, because it allows a sufficiently short experiment. The strand central part of 50 mm length was only subjected to corrosion; remaining steel surface in corrosion cell has been coated with an electric insulator (Fig. 2).

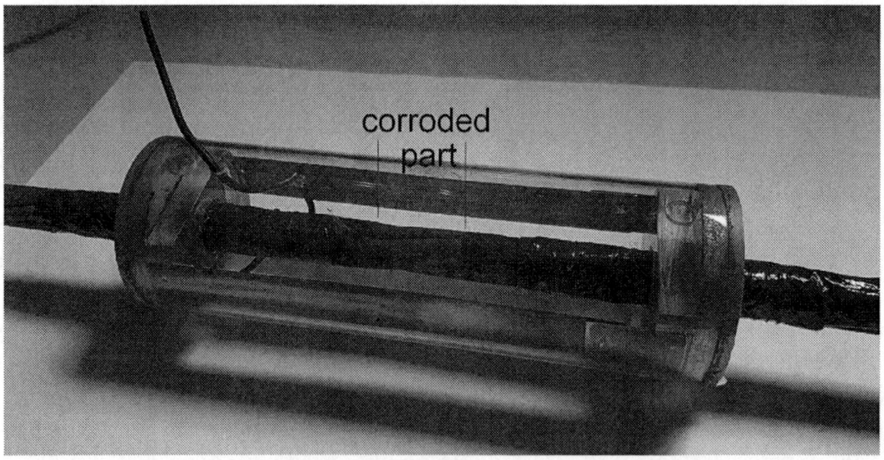

Figure 2. The strand equipped with corrosion cell.

3. Experimental part

The accelerated corrosion was intermitted and its consequences were analysed by use of linear and nonlinear methods. Measurements have been carried in four steps: before accelerated corrosion beginning and in periodical 14-daystime intervals (three steps) during the accelerated corrosion procedure (42 days). The linear and nonlinear elastic properties of the strand were interpreted from the recorded data.

3.1 Linear measurements

Impact-echo method based on the propagation of stress waves, which are generated by a mechanical impulse was used (Fig. 3). We determine the low-amplitude response of the strand and evaluate the changes in the linear material characteristics – resonance frequency/wave speed as a function of the corrosion procedure.

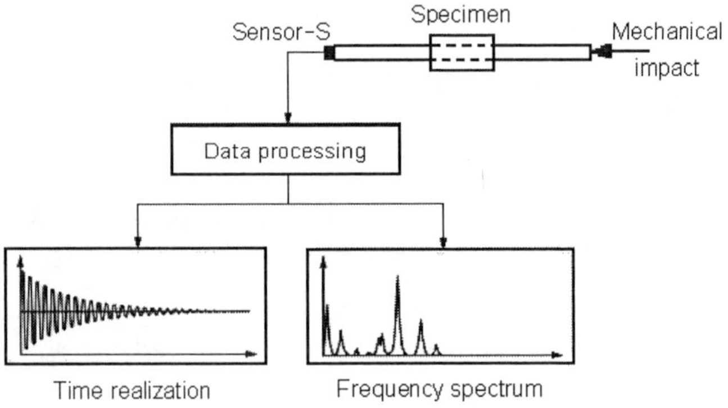

Figure 3. The schema of arrangement used in the Impact-echo method.

A short-duration mechanical impact, produced by tapping a hammer or a small steel ball against the surface of tested object, produces low-frequency stress waves that propagate into the structure and are reflected by flaws and/or external surfaces [12]. Reflected waves are recorded on the surface by a sensor in the form of a voltage signal. The resulting voltage versus time plot (time-domain realization) is digitized and fed into the memory of a computer, which subsequently carries out the frequency analysis of it. A time realization and the corresponding frequency spectrum are the results of this test. The predominant frequencies (which are represented by local maxima in the spectrum) may be associated with multiple reflections within the structure, carrying information on the structure integrity and defect localization. A mechanical impulse, provided in our case by a special hammer, and the sensor S was used to pick up the response (see Fig. 3). The measurements gave rise to frequency spectra of the specimen response to a mechanical impulse, as shown in Fig. 4.
Dominant frequencies (denoted a, b, c, d) pertain to measurement results in gradual stages of the accelerated corrosion action. The longitudinal direction predominant frequency of strand with corrosion cell, before corrosion procedure beginning (intact strand: line-a), equals 4 050 Hz. The diagram shows clearly that the predominant frequency values decrease with gradual strand corrosion. Table 1 summarizes frequencies of longitudinal oscillations for all measurements during the accelerated corrosion procedure. It is seen that the predominant frequency value, after 14 days, shifted to 3 360 Hz, which corresponds to a shift of value 690 Hz. The predominant frequency value, after 28 days, shifted to 2 520 Hz, which makes, in comparison with frequency value of intact strand, a shift of value 1 530 Hz. Final

predominant frequency shifted to 1 980 Hz, which makes a shift of value 2 070 Hz in comparison with frequency value of intact strand with corrosion cell. The experimental value of the longitudinal direction predominant frequency amounts 4 110 Hz for the intact strand without corrosion cell. This experimental value is comparable with theoretical calculated value 4 180 Hz.

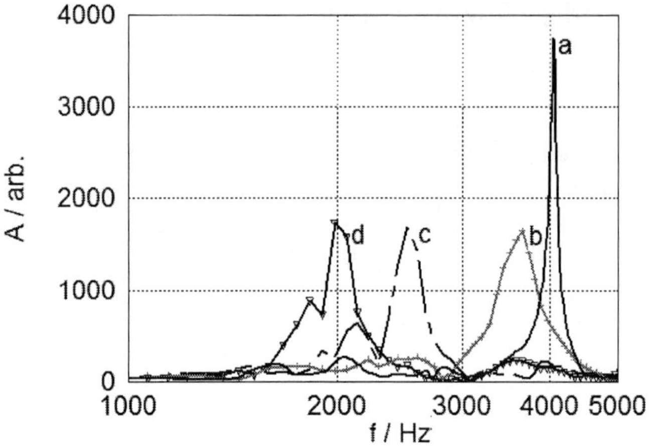

Figure 4. Response frequency spectrum of the strand with corrosion cell: a – before corrosion procedure (intact strand), b – after 14 days, c – after 28 days, d – after 42 days.

Table 1. Values of predominant frequency in dependence on corrosion stage

Corrosion stage	before corrosion (a- intact strand)	b - after 14 days	c - after 28 days	d - after 42 days
Predominant frequencies	4 050 Hz	3 360 Hz	2 520 Hz	1 980 Hz

3.2 Nonlinear measurements

Nonlinear dynamic response may manifest itself in a variety of manners, in this study, nonlinear ultrasonic techniques were applied to investigate the amplitude dependence of the resonance frequency. The used method is a nonlinear frequency domain method, also termed single mode nonlinear resonant acoustic spectroscopy (SIMONRAS), and investigates the shift of the resonance frequency as function of the resonance amplitude.

The experimental set-up and the testing of its component units have been described in detail [13, 14] and is briefly described here. The measuring apparatus consists of two principal parts, namely, a transmitting section and a receiving and measuring section. The transmitting section consists of four functional blocks: a controlled-output-level harmonic signal generator, a low-distortion 100 W power amplifier, an output low-pass filter to suppress higher harmonic components and ensure high purity of the exciting harmonic signal to exciter (denoted E in Fig. 5). The main chain of the receiving section includes an input amplifier with filters designed to minimize the receiving chain distortion and a band-pass filter amplifier. Having been amplified, the sensor (denoted S in Fig.5) output signal was fed into a Tie Pie Handyscope HS5 - XM 500 HM measuring instrument to be sampled and analyzed. For the purpose of improving the reliability and accuracy of the nonlinear experiments and minimizing the error effects the attention was focused to transmission between exciter (E) and sensor (S). A program package to control the measuring process, the data processing and evaluation makes an indispensable tool. The photo of the experimental arrangement is showed in Fig. 5.

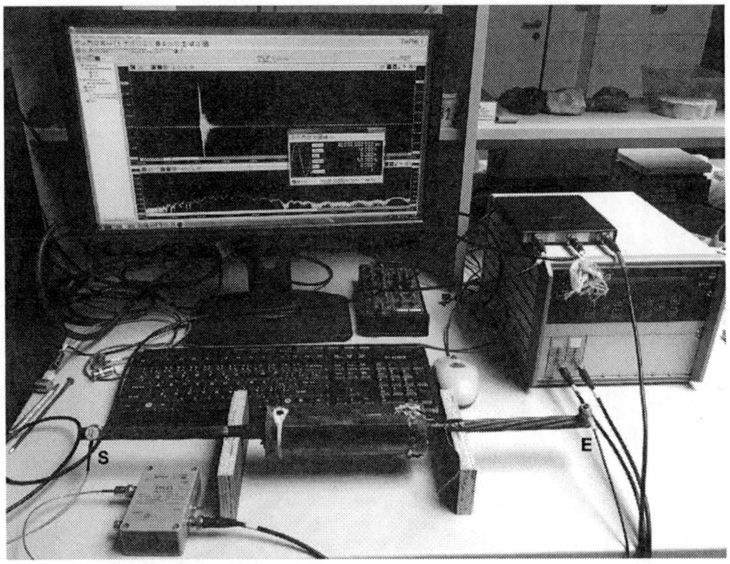

Figure 5. The photo of the experimental arrangement: E – exciter, S – sensor.

An exciting impulse, provided in our case by an electrical harmonic sinusoidal signal, was applied at the exciter point E, a piezo-electric sensor S was used to pick up the response (see Fig. 5). In order to monitor frequency peak shift as function of the amplitude, four sweeps are made at successively drive exciting voltage values (12V, 10 V, 8 V, 6 V) over the same corrosion stage. The measurements gave rise to frequency spectra of the strand response to an electrical impulse as shown in Fig. 6 through 10. Fig. 6 corresponds to the first measurement results before corrosion procedure beginning, for voltage value 12V. The predominant frequencies occur in two frequency ranges as it is shown in this response frequency spectrum.

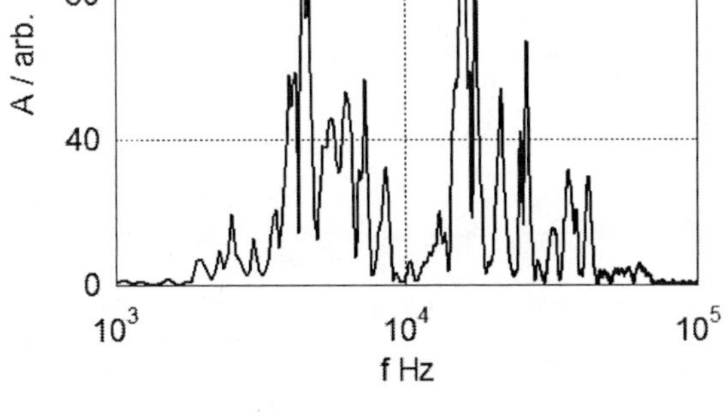

Figure 6. Response frequency spectrum corresponding to the measurement before accelerated corrosion action; the whole frequency range.

Following Fig. 7 shows detail of predominant frequencies. The first dominant frequency value equals 4 500 Hz for all exciting voltage values as is evident in graph B and the second dominant frequency equals 15 700 Hz for all exciting voltage values as it is analogically apparent in graph C.

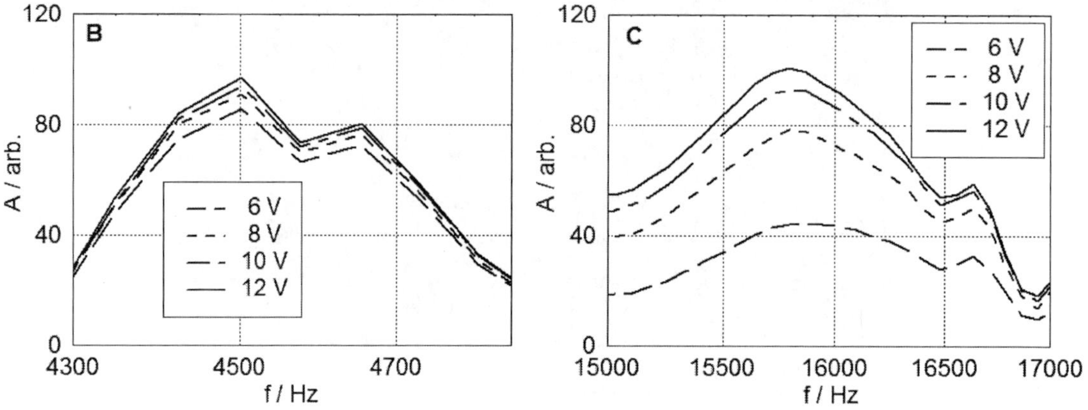

Figure 7. Details of predominant frequency ranges correspond to four exciting voltage values.

Next Fig. 8 corresponds to the second measurement; the strand was submitted to accelerated corrosion fourteen days. The predominant frequencies occur in the second frequency range only, as it is shown in diagram A. Diagram B shows the predominant frequency range in detail. It is seen that predominant frequency peaks depend on exciting voltage value. The predominant frequency value is 27 900 Hz for the highest voltage value 12 V. Dominant frequency peak values successively growth with decreasing voltage value. The dominant frequency equals 28 300 Hz for the lowest voltage value 6 V, it corresponds to shift of 400 Hz in comparison with the highest exciting voltage value 12 V.

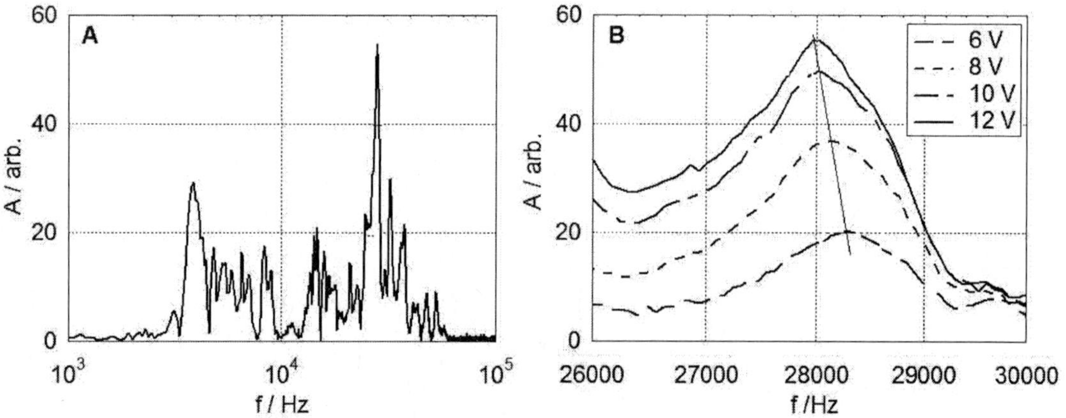

Figure 8. Response frequency spectrum corresponding to the measurement after fourteen days accelerated corrosion action: A-the whole response frequency spectrum, B-detail of predominant frequencies.

Fig. 9 analogically compares measurement results after 28 days of accelerated corrosion action. Dominant frequencies occur in the second frequency range again as it is apparent from diagram A. The predominant frequencies (diagram B) prove similar values as shown in Fig. 8, B. Dominant frequency pertains exciting voltage value 12 V is 26 800 and values of dominant frequencies growth stepwise with decreasing voltage value. Dominant frequency value 27 100 Hz corresponds the lowest exciting voltage value 6 V. The frequency shift amounts to 300 Hz in comparison with the highest voltage value 12 V.

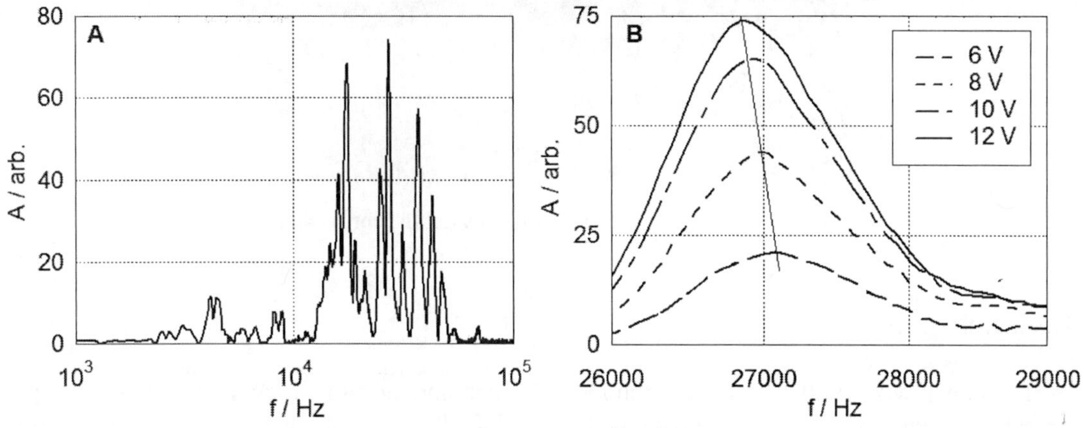

Figure 9. Response frequency spectrum corresponding to the measurement after 28 days accelerated corrosion action: A-the whole response frequency spectrum, B-detail of predominant frequencies.

Next graph in Fig. 10 (on the left) represents results of the last measurement after 42 days accelerated corrosion action. In this case the highest frequency amplitudes occur on the background level. At the end of the measurement, corrosion procedure was finished and resulting damage was visually inspected. The photo, in Fig. 10 (on the right), shows clearly consequences of accelerated corrosion action. The corroded part of the strand is covered by layer of rust and corrosion species.

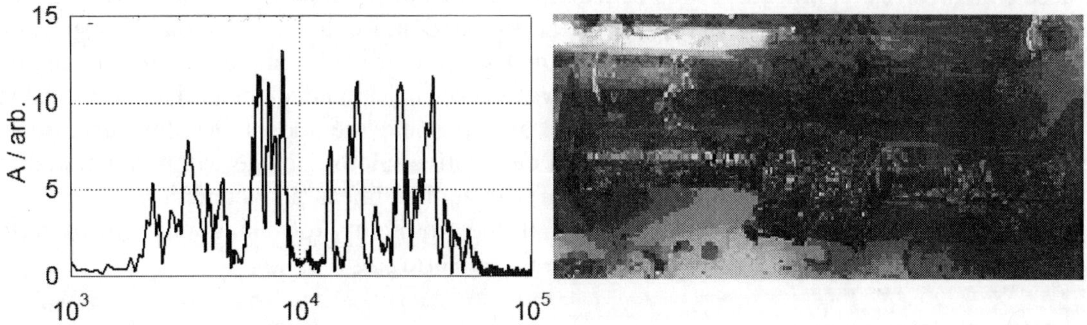

Figure 10. Response frequency spectrum corresponding to the 42 days accelerated corrosion action – on the left; photo of the strand at the end of the accelerated corrosion procedure - on the right.

The strand appearance after removal of the rustyer is showed in Fig. 11. It is clear, from this photo that all six external wires corroded completely and the central wire (in middle part which was immersed in corrosion solution) was weakened and ruptured in a consequence of corrosion.

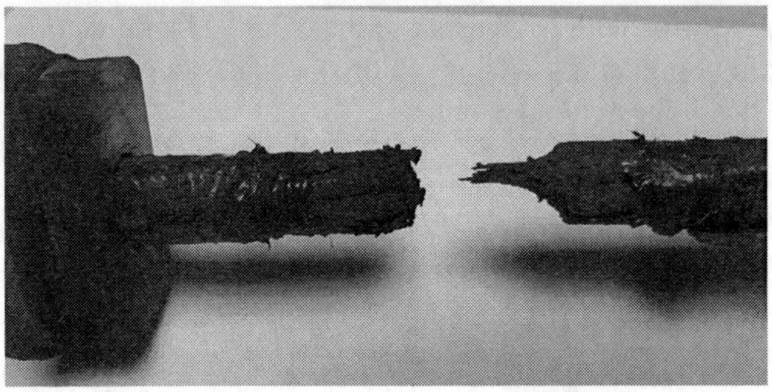

Figure 11. Ruptured strand after removal of the corrosion layer.

4. Conclusions

In this paper we investigated the changes in linear and nonlinear characteristics of a pre-stressing strand as damage is progressively induced by accelerated corrosion procedure. Measurements have been carried in four steps: before accelerated corrosion beginning and in periodical 14-daystime intervals (three steps) during the accelerated corrosion procedure (42 days). In case of the linear characteristics measurements, we determine the low-amplitude response of the strand and evaluate the changes in the linear material characteristics – resonance frequency/wave speed as a function of the corrosion procedure. Frequency spectra showed that predominant frequency values successively decreased with accelerated corrosion action. Frequency spectrum corresponds to the last measurement showed the same trend without changes which were able to characterize strand rupture due to corrosion.
In case of the nonlinear measurements, we investigated the shift of the resonance frequency as function of the exciting amplitude/exciting voltage value. Progressive damage of the strand produced the shift predominant frequencies in dependence on exciting amplitude. The accelerated corrosion procedure was ended when values of predominant frequencies (the last measurement after 42 days) appeared on background noise level. A visually inspection showed strand resulting damage: all six external wires corroded completely and the central wire was ruptured. High transmission loss was caused by strand damage due to corrosion procedure. The advantage of applied nonlinear ultrasonic method is that the increase in nonlinearity is the most sensitive indicator of cumulative micro-damage within a material in consequence of corrosion attack in comparison with applied linear characteristics.
Further studies will be performed to exploit the possibility of using our approach to monitoring of condition of strands which are built into bridge girders.

Acknowledgments

This article was produced on the side of CDV with the financial support of the Ministry of Transport within the programme of long-term conceptual development of research institutions, on the research infrastructure acquired from the Operation Programme Research and Development for Innovations (CZ.1.05/2.1.00/03.0064) and on the side of the Brno University of Technology with the support of project No. LO1408 "AdMaS UP - Advanced Materials, Structures and Technologies", awarded by Ministry of Education, Youth and Sports.

References

1. S. Ramadan, L. Gailet, C. Tessier, H. Idrissi, 'Detection of stress corrosion cracking of high-strength steel used in prestressed concrete structures by acoustic emission technique', Applied Surface Science 254 (2008), pp. 2255-2261.
2. M. Moreno, W. Morris, M. G. Alavarez, G.S. Duffo, 'Corrosion of reinforcing steel in simulated concrete pore solutions: Effect of carbonation and chloride content', Corrosion Science, 46 (2004), pp. 2681-2699.
3. B. Huet, V. L.Hostis, F. Miserque, H. Idrissi, 'Electrochemical behavior of mild steel in concrete: Influence of pH and carbonate content of concrete pore solution', Electrochimica Acta 51 (2005), pp. 172-180.
4. H. Idrissi, A. Liman, 'Study and characterization by acoustic emission and electrochemical measurements of concrete deterioration caused by reinforcement steel corrosion', NDT & E International 36 (2003), pp. 563-569.
5. Z. Liu, S. Liu, B. Wu, Y. Zhang, C. He, 'Propagation characteristics of high order longitudinal modes in steel strands and their applications', Acta Mechanica Solida Sinica, 21 (2008), pp. 573-579.
6. A. Yonezu, H. Cho, M. Takemoto, 'Monitoring of stress corrosion cracking in stainless steel weldments by acoustic and electrochemical measurements', Measurement Science and Technology 17, (2006), pp. 2447-2454.
7. M. M. EI Rayes, H. S. Abdo, K. A. Khalil, 'Erosion - Corrosion of Cermet Coating', International Journal of Electrochemical Science 8 (2013), pp. 1117-1137.
8. S. Q. Wang, D. K. Zhang, D. G. Wang, L. M. Xu, S. R. Ge, 'Stress Corrosion Behaviors of Steel Wires in Coalmine under Different Corrosive Mediums', International Journal of Electrochemical Science 7 (2012), pp. 7376-7389.
9. M. Kořenská, L. Pazdera, K. Pospisil, J. Stryk, P. Vyroubal, 'Detection of Reinforcement Corrosion in Prestressed Concrete Girdes', Proc. of The 8th International Conference of the Slovenian Society for Non-Destructive Testing, Portorož, Slovenia, 2005, pp. 317-322.
10. M. Kořenská, M. Manychová, L. Pazdera, K.Pospíšil, 'Monitoring of the Damage Evolution in Reinforced Concrete Girder by Means of Nonlinear Elastic Wave Spectroscopy', Transactions on Transport Science, 5 (2012), pp. 195-204.
11. S. Caré, Q. T. Nguyen, V. L. Hostis, Y. Berthaud, 'Mechanical properties of the rust layer induced by impressed current method in reinforced mortar', Cement and Concrete Research 38 (2008), pp. 1079-1091.
12. O. Abraham, C. Leonard, P. Cote, 'Time frequency Analysis of Impact-echo Signals: Numering Modeling and Experimental Validation', ACI Materials J., 97 (2000), pp. 645-657.
13. M. Manychova,, 'Measuring setup for the non-linear ultrasonic spectroscopy method', Proc. of The 37th International Conference Defektoskopie 2007, Prague, 2007, pp. 147-152.

14. M. Korenska, L. Pazdera, M. Matysik, 'Testing of Pickups for Single-exciting-signal Nonlinear Acoustic Spectroscopy', Proc. of International Conference Physical and Material Engineering 2006, Bratislava, Slovak Republic 2006, pp. 70-7

Optimization of Inclined Angle Magnetic Sensor Probe to Improve Detection Performance of Corrosion near the Ground using Extremely Low-frequency Eddy Current Testing

Shunki WAKABAYASHI, Minoru HAYASHI Taisuke SAITOH,
Kenji SAKAI, Toshihiko KIWA, Keiji TSUKADA

Okayama University, Okayama 700-8530, Japan
E-mail:pf286cwm@s.okayama-u.ac.jp

Abstract
As the aging of steel pipes used for sign pole is caused by the corrosion near ground, a quantitative evaluation of the corrosion location and thinning rate from the ground surface is required. Therefore, we have proposed a magnetic nondestructive testing for the steel pipes using two magnetic sensors, and it was possible to inspect the underground corrosion defects. However, in order to improve the accuracy of estimating the thinning rate and position of underground corrosion, the optimization of the inclination angle of the magnetic sensor probe and the distance between two magnetic sensors has been required. Therefore, in order to extract the signal change caused by the thinning rate of the test sample and the measurement distance, the distance dependence which is the relationship between the signal intensity and the measurement distance was measured and the measured value was normalized by the signal intensity measured at uncorroded area. As a result, the difference between the signal intensity at the uncorroded part and that near the corrosion defect could be clearly extracted. Furthermore, the rate of change of signal intensity increased by changing the tilt angle of probe and the distance between two sensors. Using the optimized magnetic sensor probe and applying the normalized distance dependence, we succeeded in the quantitative evaluation of the thinning rate and position of corrosion defects.
Keywords: Nondestructive testing, ELECT, corrosion near the ground, Magnetic spectrum curve

1. Introduction

The aging of steel pipes used for lighting and sign poles that are a part of social infrastructure is becoming a problem. This aging is often caused by the corrosion of steel pipes generated near the ground which is located several tens of millimeters from the ground line. The accident that the structure collapses has been reported due to the growth of ground corrosion defects [1]. Currently, a general maintenance inspection of ground corrosion defect is a visual inspection performed by excavating the ground. However, it is difficult to detect the corrosion defects near the ground quickly and quantitatively with such an inspection method. Based on the extremely low frequency eddy current testing (ELECT) [2-4], we have developed the integrated magnetic sensor probe with two tilted sensors in order to detect the corrosion near the ground from the ground surface [2, 5, 6]. Using this probe, a correlation between the signal attenuation due to the corrosion near ground and the thinning rate was obtained [2, 6]. In order to improve the accuracy of estimation for the thinning rate and position of underground corrosion, the optimization of the inclination angle of the magnetic sensor probe and the distance between two sensors has been required.

To improve the accuracy of corrosion position, it is necessary to increase the rate of change in distance dependence which is the relationship between the signal intensity and the measurement distance from the corrosion defect. In this sutdy, the distance dependence was measured using the test sample that assumes corrosion defect. Using the distance dependence normalaized at the uncorroded part, we attempted to clearly detect the difference between the signal intensity at the uncorroded area and the corrosion defect. The increase of the rate of change in the signal intensity was also investigated by optimizing the inclination angle (30°, 40°, and 45°). In addition, we developed the magnetic sensor probe with two sensors and optimized the distance between two sensors so that the rate of change in the signal intensity increases. Based on these results, the normalized distance dependence measured by the optimized magnetic sensor probe was calculated, and examined the method that can quantitatively evaluate the thinning rate and position of corrosion defects.

2. Measurement

The schematic of measuring system and test sample is shown in Fig 1. The system consisted of a current source, a function generator, a magnetic sensor probe, an amplifier, a lock-in detector, and a PC. The magnetic sensor probe consisted of an anisotropic magneto resistive (AMR) sensor, an induction coil and a cancellation coil. The test sample was a steel plate with three types of defects; 60 mm in wide and 1, 2, and 3 mm in depth assuming corrosion defects. The plate thickness of the uncorroded part was 4 mm. The test sample dimensions were 200 mm × 350 mm and the material was SS400. The applied frequency was 3 and 50 Hz, and the magnetic field vector of each frequency was measured. Furthermore, the intensity of the differential magnetic field vector was calculated by subtracting the magnetic field vector of 3 Hz from that of 50 Hz. Using the intensity of this differential magnetic field vector, it is possible to extract the signal attenuation due to the corrosion.
Measurements were performed at 10 mm intervals from the thinning zone. The lift off which is the distance between the sample and the edge of probe (see Fig. 1) was 10 mm and the applied current was 200 mA. Additionally, the distance dependence which is the relationship between the signal intensity and measurement distance from the thinning zone was obtained.

3. Result

Figure 2 shows the distance dependence before and after normalization. The values of Fig. 2(b) shows the normalized one by the signal measured at the uncorroded part sufficiently away from the corrosion defect. Regardless of the normalization, the differential intensity decreased as the defect depth increased.

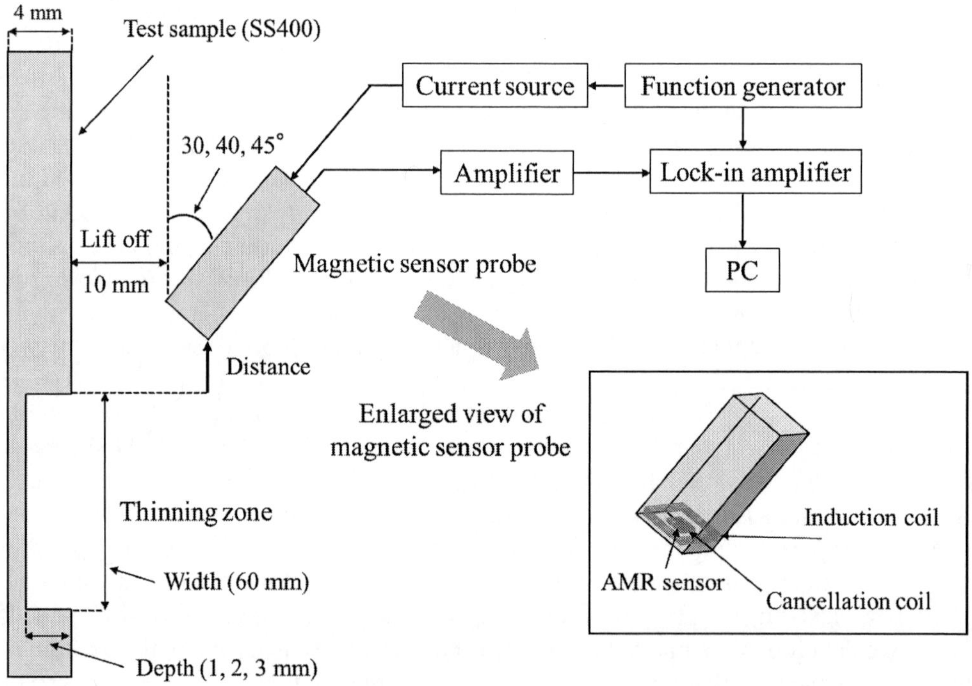

Fig. 1. Configuration of test sample and measuring system.

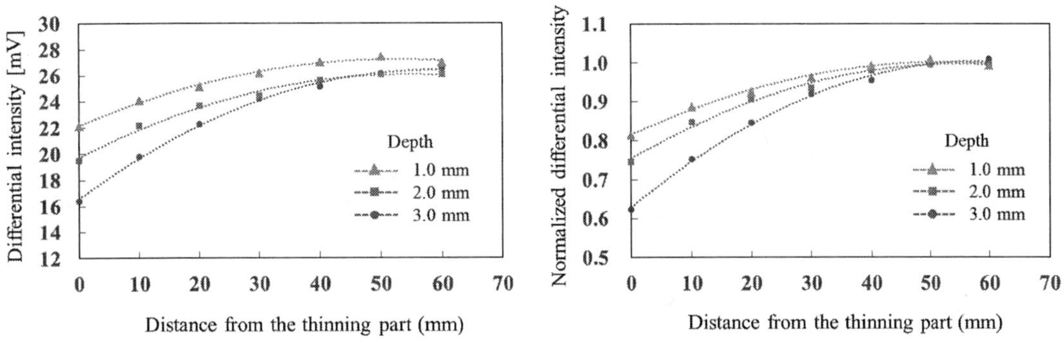

(a) Before normalization

(b) After normalization

Fig. 2. Distance dependence before and after normalization.

However, when the measured signal was normalized, the attenuation of the differential intensity for the uncorroded part was quantitatively evaluated. Furthermore, the normalized differential intensity approached 1.0 at a measurement distance of more than 50 mm (Fig. 2(b)). This means that the differential intensity was similar to that of the uncorroded part and was saturated. These results indicate that the normalized differential intensity enabled to quantitatively extract the difference of signal intensity between the uncorroded area and the corrosion defect.

Then, the tilt angle of the magnetic sensor probe was optimized to improve the detection performance of corrosion defects. The differential signal was measured using three types of tilt angle; 30°, 40° and 45°, and the rate of change of the differential intensity was examined. The test sample used had a defect with the depth of 3 mm, and the measurement conditions were the same as those for the measurement of the distance dependence.

Figure 3 shows the distance dependence of each tilt angle. When the tilt angle was 45°, the differential intensity was approximately 0.6, which was the smallest at the measurement distance of 0 mm (Fig. 3(c)). However, when the measurement distance increased, the differential intensity converged to the same level as that of the uncorroded part at a position of approximately 40 mm. Moreover, when the inclination angle was reduced to 30° or 40°, the normalized differential intensity was approximately 0.9 which is less than the uncorroded part even at the position of 40 mm (Figs. 3(a), (b)). Thus, it was found that the differential intensity changed compared to that at uncorroded part at the tilt angle of 30° and 40° even if the measurement distance was increased. In this study, tilt angle of 40° is considered to be optimum because the differential intensity greatly decreased when the measurement distance was between 0 and 30 mm, and the differential intensity also changed when the measurement distance increased more than 30 mm.

In Fig. 4, the magnetic sensor probe with two tilt sensors which were fixed at 40° is depicted. The distance between the two sensors can be changed, and the distance between the sensors was optimized to estimate the position of the corrosion defect. The distance dependence of four sensor distances (23, 30, 35, and 40 mm) were measured. The defect depth of the test sample used was 3 mm, and the measurement conditions were the same as those for the tilt angle optimization.

Figure 5 shows the distance dependences of each sensor distance. As the distance between sensors decreased, the rate of change of the differential intensity for sensor 2 tended to increase. In the case of the closest sensor distance of 23 mm, the differential intensity of sensor 2 was 0.85 at the measurement distance of 0 mm and this differential intensity was the smallest (Fig. 5(d)). Furthermore, the attenuation of the differential intensity can be confirmed up to a measurement distance between 30 and 40 mm. Based on these results, the distance between the sensors of 23 mm was selected.

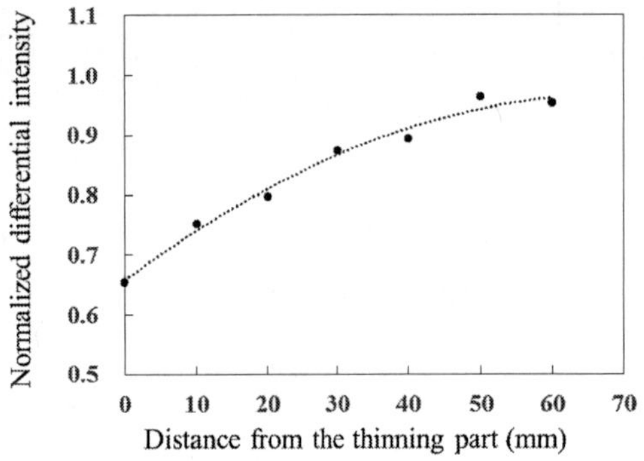

(a) Inclination angle 30°

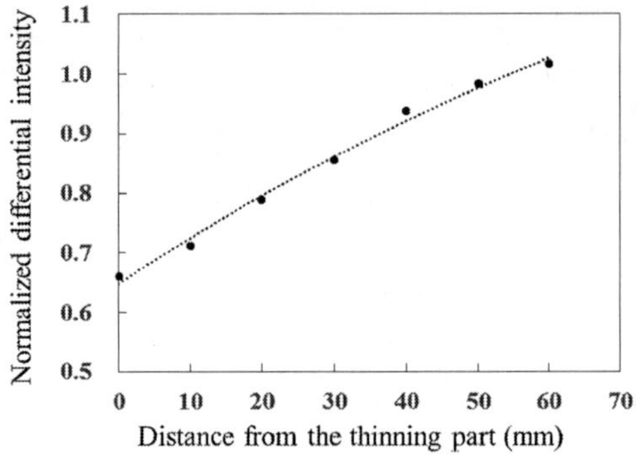

(b) Inclination angle 40°

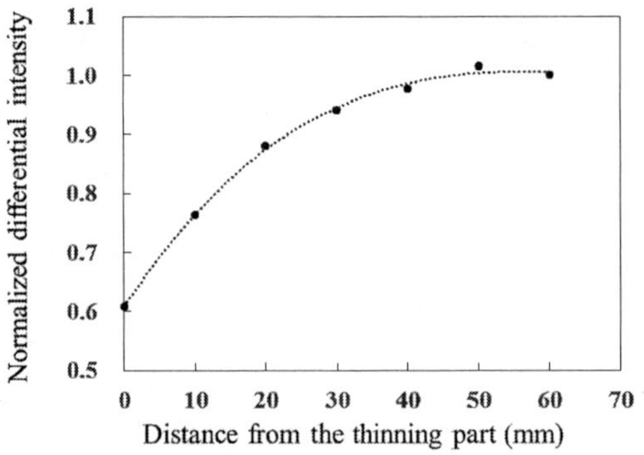

(c) Inclination angle 45°

Fig. 3. Distance dependence for each inclined angle.

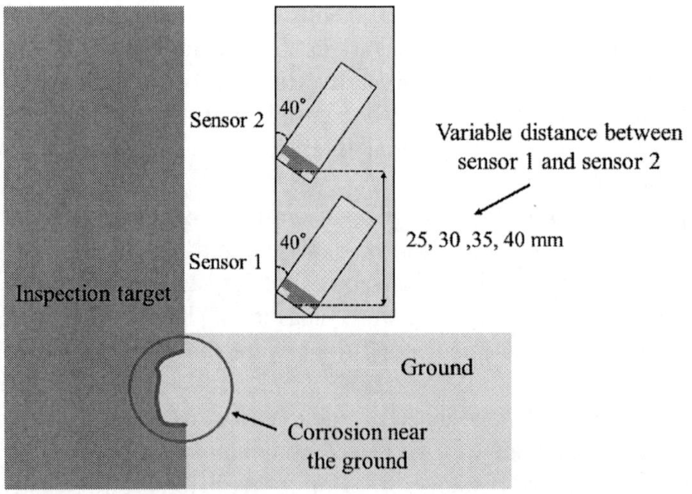

Fig. 4. Configuration of the developed variable magnetic sensor probe.

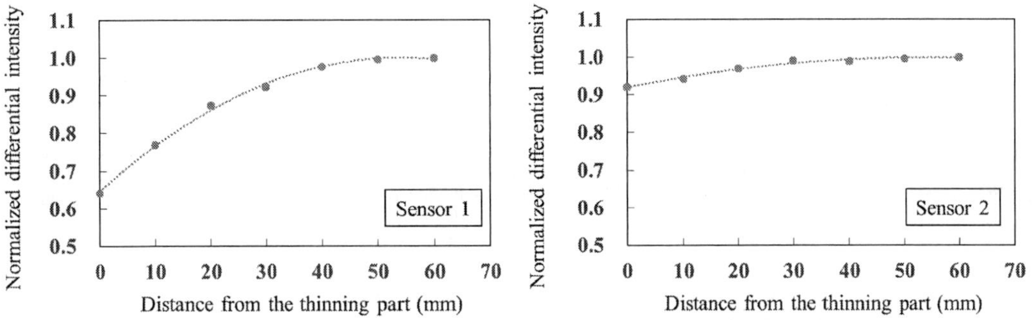

(a) Distance dependence of 40 mm distance between two sensors

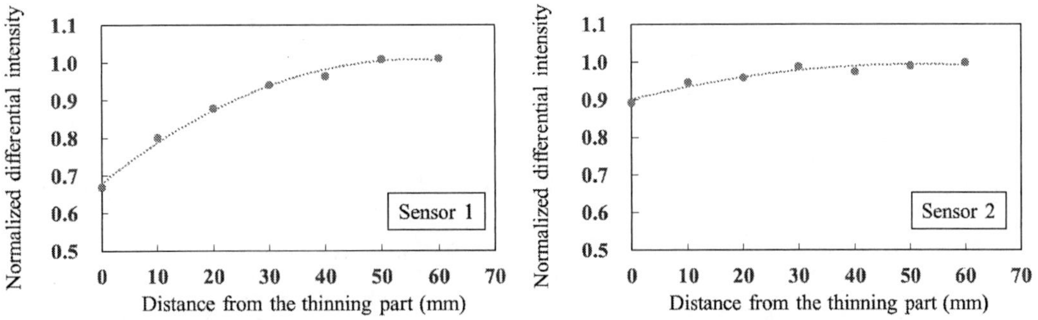

(b) Distance dependence of 35 mm distance between two sensors

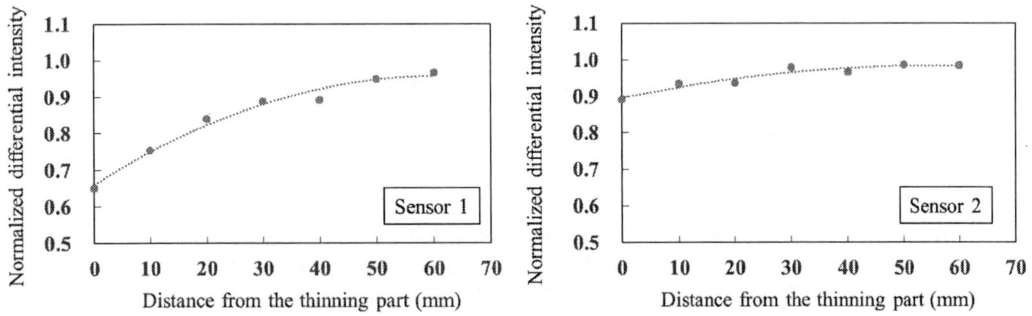

(c) Distance dependence of 30 mm distance between two sensors

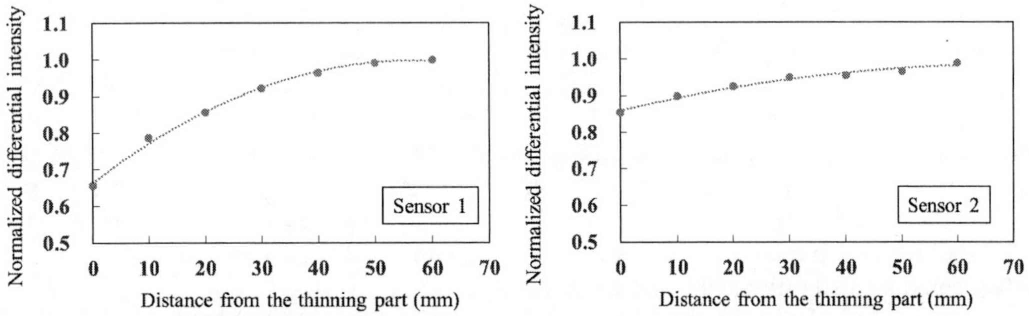

(d) Distance dependence of 25 mm distance between two sensors

Fig. 5. Distance dependence of each distance between two sensors.

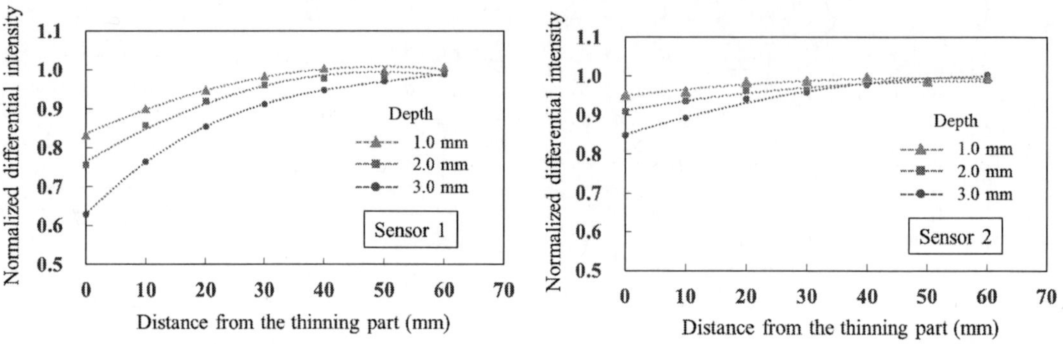

Fig. 6. Distance dependence of samples with different thinning rates.

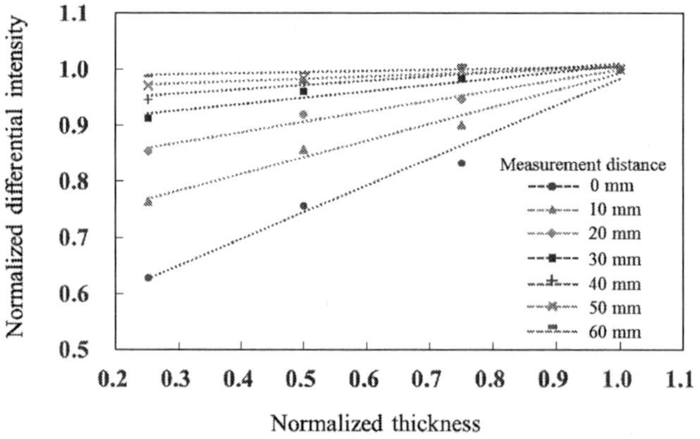

Fig. 7. Relationship between normalized differential intensity and plate thickness.

Considering the obtained results, the magnetic sensor probe with the inclination angle of 40° and the 23 mm-distance between sensors was developed. The distance dependence obtained by test samples with different thinning rate using the developed magnetic sensor probe is shown in Fig. 6. In the distance dependence of sensor 1, it was possible to determine the defect depths up to the measurement distance near 50 mm. Moreover, in the distance dependence of the sensor 2, it was possible to determine the defect depths up to the measurement distance of approximately 30 mm. Since the differential intensity changed depending on the thinning rate, a quantitative evaluation of the thinning rate is possible. Furthermore, as the differential intensity obtained by the sensors 1 and 2 changed due to the measurement distance, the position of corrosion defects could be estimated quantitatively by using the distance dependence of the test sample as a calibration curve. For example, if the intensity of sensor 1 was approximately 0.75, and the intensity of sensor 2 which is 23 mm away from the sensor 1 was approximately 0.92, the defect depth could be estimated as 3 mm. In addition, the position of the corrosion defect could be also estimated from the sensor 1.

Next, the rate of plate thickness was calculated from the normalized differential intensity. Figure 7 shows the relationship between the differential intensity at the measurement distance of the sensor 1 and the thinning rate of the plate thickness. As the measurement distance decreased, the rate of change of the differential intensity increased. In the case of the measurement distance of 0 mm, when the differential intensity was approximately 0.75, the plate thickness was calculated as approximately 0.5 for the uncorroded part. When the differential intensity was approximately 0.65, the plate thickness was calculated in the same way as approximately 0.25 at the uncorroded part. Therefore, the rate of plate thickness can be also possible to calculate quantitatively.

5. Conclusion

In this study, the distance dependence which is the relationship between the signal intensity and the defect position from the sensor was measured and the obtained data was normalized with the signal intensity measured at the uncorroded area. As a result, it was possible to clearly extract the difference between the signal intensity at the uncorroded area and the corrosion defect. The inclination angle of the magnetic sensor probe and the distance between two sensors were optimized by measuring a test sample, and the corrosion defects located at deep position was detected. In addition, using the optimized magnetic sensor probe with two sensors, it was possible to estimate the thinning rate and position of defect. It is concluded that from the obtained results that the corrosion defects near the ground can be quantitatively evaluated using the normalized distance dependence and the relationship between differential intensity and plate thickness.

Acknowledgments

This research was conducted by the Cross-ministerial Strategic Innovation Creation Promotion Program (SIP).

References

1. T. Tamakoshi, M. Hoshino, A. Ichikawa, "Survey on Deterioration and Damage of Road Attachment Facilities", Technical Note of National Institute for Land and Infrastructure Management, No.685, 2012.
2. K. Tsukada Y. Haga, K. Morita, N. Song, K. Sakai, T. Kiwa, and W. Cheng, "Detection of inner corrosion of steel construction using magnetic resistance sensor and magnetic spectroscopy analysis", IEEE Trans. Magn., vol. 52, no. 7, 6201504, 2016.
3. G. Y. Tian, A. Sophian, "Defect classification using a new feature for pulsed eddy current sensors", NDT&E Int., vol. 38 pp. 77–82, 2005.
4. T. Tomioka, T. Goda, K. Sakai, T. Kiwa, K. Tsukada, "Imaging of internal corrosion of steel structures using an extremely low-frequency eddy-current testing method", APCND, 2017
5. Y. He, M. Pan, F. Luo, and G. Tian, "Pulsed eddy current imaging and frequency spectrum analysis for hidden defect nondestructive testing and evaluation", NDT & E Int., vol. 44, no. 4, pp. 344–352, 2011.
6. K. Tsukada, T. Tomioka, S. Wakabayashi, K. Sakai, T. Kiwa, "Magnetic detection of steel corrosion at a buried position near the ground level using a magnetic resistance sensor", IEEE Trans. Magn., vol. 54, no. 11, 6202104 2018.
7. S. Wakabayashi, T. Tomioka, K. Sakai, T. Kiwa, K. Tsukada, "Improvement in the detection performance of extremely low-frequency eddy current testing for application in underground steel corrosion detection", APWSHM, November 2018

Magnetic State Observer in NDT and SHM Studies

Miroslaw WITOS [1], Marek SZYMANSKI [1], Milan ORAVEC [2], Karol BOGUCKI [3]

[1] Air Force Institute of Technology, Warsaw, Poland
Phone: +48 261851353, Fax: +48 261851612, e-mail: witosm@itwl.pl, marek.szymanski@itwl.pl
[2] Technical University of Košice (TUKE), Košice, Slovakia, E-mail: oravec.milan@cetrum.sk
3 Military Aviation Works No 4 JSC, Warsaw, Poland, E-mail: boguckikarol1217@gmail.com

Abstract
The article presents the project for the use of magnetic magnetometers in the 0 - 500 Hz band for remote, non-contact monitoring of object operation and assessment of the technical condition of ferromagnetic elements and systems. At the beginning, the theoretical foundations of passive magnetic tests and applied magnetometers (multi channels 3 D system) have been presented. Next, selected results of laboratory tests and applied measurement data analysis methods are presented. Then selected results of tests carried out on the real object are presented, among others Kaplan turbine and turbine jet engine. Finally, discussing the results of research, the strengths and weaknesses of the magnetic state observer were pointed out. It has been experimentally demonstrated that a remote magnetic state observer provides qualitative and quantitative diagnostic information that can be used in non-destructive testing (NDT) and structural health monitoring (SHM).

Keywords: fatigue of material, non-destructive testing (NDT), structural health monitoring (SHM), magnetometer, measurement, signal analysis

1. Introduction

During the operation of machines and devices, the wear potential [1] of individual elements and systems. In order to maintain the airworthiness of technical facilities:
- work parameters are monitored;
- service is performed, including non-destructive tests, resulting from the adopted operating system and the results of statistical analysis of adverse events. There are three basic mining systems [2]:
  a) planned prevention with known time between overhauls (TBO) and resources;
  b) technical condition;
  c) the reliability requested.

Most industrial facilities and aircraft engines are operated according to their technical condition or according to planned prevention - useful life period pre-determined by the constructor and manufacturer based of construction assumptions verified during factory and qualification tests. The technical object's lifetime is extended based on:
- values of statistical indicators of operational reliability and safety determined for the population of objects which are also affected by human factors present in the system of design, production, training, operation, renovation and logistic protection [3 - 6];
- new technical and technological possibilities, including modern control and measuring apparatus and diagnostic methods as well as new technologies implemented for the operation, repair and quality control of spare parts production.

Along with increasing the reliability of the diagnosis and the forecast horizon, operational safety increases and operating costs decrease, which was used in the operating system according to technical condition.
Reliable determination of the technical condition of critical elements and forecasting the horizon of their safe operation requires:
- good knowledge of the test object, including construction and material data, which are most often inaccessible to the user and the diagnostician;
- knowledge of the history and effects of real effort;

- proper selection of the state observer [2, 7], based the relationships between the technical condition of the test object and its elements with:
  a) measurable physical and mechanical parameters,
  b) available measurement methods,
  c) costs of prevention;
- basic knowledge of signal theory and algorithms for measuring data pre-processing;
- verified knowledge of experts or knowledge obtained from machine learning, necessary to aggregate the characteristics of measurement signals and to separate estimators and diagnostic criteria (covert relations with degradation of the chemical composition and structure of the material [8, 9]).

The effectiveness of preventive actions, described by increasing the probabilities of diagnosis (POD) and decreasing statistical indicators of faults and failures, is the resultant of:
- the expected diagnostic symptoms (knowledge about the process and accompanying phenomena);
- the choice of observation methods: **E**lectro-**M**agnetic **I**nterference (EMI), **C**ondition **M**onitoring (CM), **N**on-**D**estructive **T**esting (NDT), **S**tructural **H**ealth **M**onitoring (SHM), **P**rognostic and **H**ealth **M**anagement (PHM);
- the quality of the measurement path, software quality and diagnostic criteria used for numerical analysis.

The article presents the possibility of obtaining new diagnostic information about the current operating conditions of the technical object and the effects of the unknown history of straining ferromagnetic material based on the cheap, **P**assive **M**agnetic **S**tate **O**bserver (PMSO). The observer combines the functions: control of low-frequency EMI, method of NDT, incomplete observer of CM, method of SHM. Measurement and diagnosis results can be used to reliably predict technical condition in PHM systems. The topic discussed are illustrated by examples of research results on industrial and aviation facilities.

2. Research objects

Two groups of technical objects will be considered - presented in the Table 1, with different: design features, operating conditions, material degradation processes and magnetic properties. Their main features are expected high reliability and operational safety. In the general case, the material of the test object is exposed to a complex state of stress and deformation caused by external forces which generates complex three-dimensional (3D) states of stress and strain and material fatigue: **L**ow **C**ycle **F**atigue (LCF), **H**igh **C**ycle **F**atigue (HCF), **V**ery **H**igh **C**ycles **F**atigue (VHCF), **T**hermo-**M**echanical **F**atigue (TMF), with negative impact [8]:
- erosion,
- corrosion,
- foreign matter,
- pulsation of air, steam or water flow,
- rotor imbalance and misalignment,
- dynamics of transient and short-term resonance impacts.

3. Theoretical basis for diagnosing DUT

The complex process of effort, material degradation and risk of emergency damage to critical elements can be described by simple **M**ultiple **I**nput, **S**ingle **O**utput (MISO) model including:
- geometric features of **D**evice **U**nder **T**esting (DUT),
- mechanical properties of the DUT material and its surroundings,

Table 1. Specificity of DUT

DUT	Example	Specific speed	Casing	Internal components
Industrial	Water, steam and gas turbine	Low to medium (up to 3000 rpm)	Ferromagnetic	Ferromagnetic, few paramagnetic
	Gears			Ferromagnetic
	Kinematic system		---	Ferromagnetic
	Transformer	---	Soft ferromagnetic	Paramagnetic winding, soft ferromagnetic core, $\mathbf{B} = f(I_{load})$
	AC and DC generator	Low to medium (up to 3000 rpm)		
Air	Aircraft (airplanes, helicopters	---	Paramagnetic (composite or aluminium alloys)	Paramagnetic, few ferromagnetic
	Turbine engine	High (up to 40000 rpm)	Paramagnetic	Paramagnetic, few ferromagnetic
	Gears	Low, medium and high	Paramagnetic	Ferromagnetic or paramagnetic
	Kinematic system	---	---	Paramagnetic, few ferromagnetic
	Landing gear	---	Ferromagnetic or paramagnetic	Ferromagnetic

- the actual load spectrum and severity of adverse phenomena (e.g.: erosion, corrosion, human errors),
- design working conditions and accepted strength criteria.

Most of the above data is unknown to the DUT user and diagnostician - the object being diagnosed is a "black box". The implicit influence of input signals, however, changes the technical condition and energy status of DUT material, including:

- **operational stress** σ_e (aperiodic component σ_a and dynamic component σ_d), which in general are generated by external forces F_z: centrifugal force F_r, force of gravity F_g, bending moment M_g and torque M_g and apparent strength - the force of inertia F_g. The state of operational stress at a given point of the critical element is described by the relation (1).

$$\begin{bmatrix} F_B \\ F_g \\ F_r \\ M_g \\ M_s \end{bmatrix} \rightarrow \sigma_e = \begin{bmatrix} \sigma_x & \tau_{xy} & \tau_{xz} \\ \tau_{yx} & \sigma_y & \tau_{yz} \\ \tau_{zx} & \tau_{zy} & \sigma_z \end{bmatrix} \quad (1)$$

- **residual stress** σ_{rp} deliberately introduced into the surface material of the critical element at the production stage to increase fatigue strength.
- **residual stress** σ_{rh} induced in the material of the critical element by the history of its operation, also occurring when the machine is at a standstill
- **total state of material stress** σ_T

$$\sigma_T = \sigma_e + \sigma_{rp} + \sigma_{rh} \quad (2)$$

- **total state of material strain** including elastic ε_s and plastic ε_p deformations

$$\sigma_T \to \varepsilon_s + \varepsilon_p = \begin{bmatrix} \varepsilon_x & \gamma_{xy} & \gamma_{xz} \\ \gamma_{yx} & \varepsilon_y & \gamma_{yz} \\ \gamma_{zx} & \gamma_{zy} & \varepsilon_z \end{bmatrix} \tag{3}$$

- **magnetization M** which has an induced component $\mathbf{M}_i$ by operational extortion and the residual component $\mathbf{M}_r$ present in the material after the extinction disappears, as described in equation (4).

$$\mathbf{M} = \mathbf{M}_i + \mathbf{M}_r \tag{4}$$

3.1. Magnetization

Magnetization - a directly immeasurable parameter describing the distribution of the magnetic field inside the ferromagnetic or metastable paramagnetic DUT, including [10 -14]:
- object geometry (demagnetization tensor);
- initial magnetic properties of the material resulting from the chemical composition and physical structure of the material containing: non-metallic inclusions and structural defects as well as stress state $\sigma_I, \sigma_{II}, \sigma_{III}$ respectively on the nano, micro and macro scale (the size of the iron atom in the allotropic alpha version Fe_α is 290 pm);
- unknown operation history and current DUT operating conditions,
- magneto-mechanical effects including magneto-mechanical damping (influence of internal friction and internal energy loss on heat dissipation);

was used for remote non-contact diagnostics of a technical object by means of a PMOS.
Magnetization can be accomplished by 3 independent variables [10]:
- **external magnetic field** with intensity **H**, which has been used in recording devices and magnetic information carriers for over 100 years;
- **stress state** σ_{ij}, which is used in applications based on magneto-mechanical effects i.e. the metal magnetic method;
- **temperature changes** ΔT, which is used, among others, for thermomagnetic and geophysical treatment.

The non-linear effect of these variables on the magnetization is described in relation (5)

$$\mathbf{M} = \mathbf{M}_i + \mathbf{M}_r = f(\mathbf{H}, \sigma_{ij}, \Delta T) = (1 + k_H)(1 + k_\sigma)(1 + k_{\Delta T})\mathbf{M}_0 \tag{5}$$

where: $\mathbf{M}_o$ - initial magnetization of the element; $k_H = k_H(t)$, $k_\sigma = k_\sigma(t)$, $k_{\Delta T} = k_{\Delta T}(t)$ - influence functions described by the physical properties of the material, among others: magnetic permeability μ_m, linear magnetostriction λ_i (longitudinal and transverse), coercivityH_c, electric resistivity ρ_e, coefficient of thermal expansion α_T, thermal conductivity coefficient λ_T.

Equation (5) can be presented in a simplified additive form (6) after ignoring the effects of multiple internal couplings in the material.

$$\mathbf{M} = \mathbf{M}_i + \mathbf{M}_r \approx \mathbf{M}_0 + \mathbf{\Delta M}_H(t) + \mathbf{\Delta M}_\sigma(t) + \mathbf{\Delta M}_{\Delta T}(t) \tag{6}$$

Magnetization occurs only in DUT material, which prevents its direct observation by non-destructive methods [15]. For remote evaluation of parameters and magnetization distribution $\mathbf{M} = \mathbf{M}(x, y, x, t)$ in DUT uses the definition of a magnetic field with equivalent magnetic field strength $\mathbf{H}_{eq}$ in a material medium with magnetic permeability $\mu_m(\mathbf{H}_{eq})$ and (Figure 1) [12 - 13]:
- magnetic field features:

a) magnetic field lines tangent to the magnetic induction vector **B** they create closed contours,
b) magnetic field lines penetrate the border separating two media with different magnetic properties (e.g. ferromagnetic DUT - air gap - paramagnetic fuselage-air between DUT and magnetometer) bending on the border;

- boundary conditions on the border of two media with different magnetic properties, described by the relations resulting from Gauss's law and Amper's law.

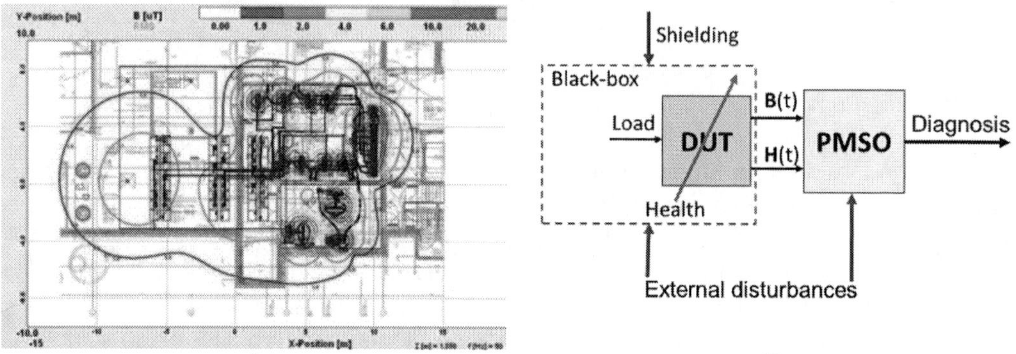

a) b)

Figure 1. Electromagnetic interference (EMI) issue: measurement and imaging of 50 Hz magnetic field near the transformer and interpretation of results: a) the field penetrates the walls of the transformation station building (remote diagnostics are possible); b) **diagnosis** - changes in magnetic field parameters are caused by: 1) a change in the transformer load or / and 2) a change in its technical condition (health) or / and 3) a change in shielding parameters or / and 4) the effect of external disturbances.

The magnetization process is described by the formula (7), where α_{ij} is the medium magnetization tensor.

$$\begin{bmatrix} M_x \\ M_y \\ M_z \end{bmatrix} = \begin{bmatrix} \alpha_{xx} & \alpha_{xy} & \alpha_{xz} \\ \alpha_{yx} & \alpha_{yy} & \alpha_{yz} \\ \alpha_{zx} & \alpha_{zy} & \alpha_{zz} \end{bmatrix} \cdot \begin{bmatrix} H_x \\ H_y \\ H_z \end{bmatrix} \quad (7)$$

In general case **M** and **H** have different directions. For paramagnetic and diamagnetic, vectors **M** and **H** are parallel. For ferromagnetic, there is no unambiguous relationship between **M** and **H** – magnetization depends on history of field strength changes and magnetic hysteresis. As a result, $\alpha_{ij} \neq \alpha_{ji}$ and the ferromagnetic stores magnetic information. The parameters of recording and resistance of magnetic information stored in the ferromagnetic describe two variables. These are (Figure 2) [16]:

- loop squareness

$$S = \frac{M_r}{M_s} \quad (8)$$

- remenent squareness

$$S^* = 1 - \frac{M_r}{H_c} \frac{1}{\left.\frac{dM_H}{dH}\right|_{H=H_c}} \quad (9)$$

where: H_c - coercivity; M_r - saturation remanence; M_s - the medium saturation magnetization; $\left.\frac{dM_H}{dH}\right|_{H=H_c}$ - magnetic susceptibility of the material for $H = H_c$.

The hysteresis loop field of $\mathbf{M(H)}$ and coercivity reflect the effect of internal friction of the material.

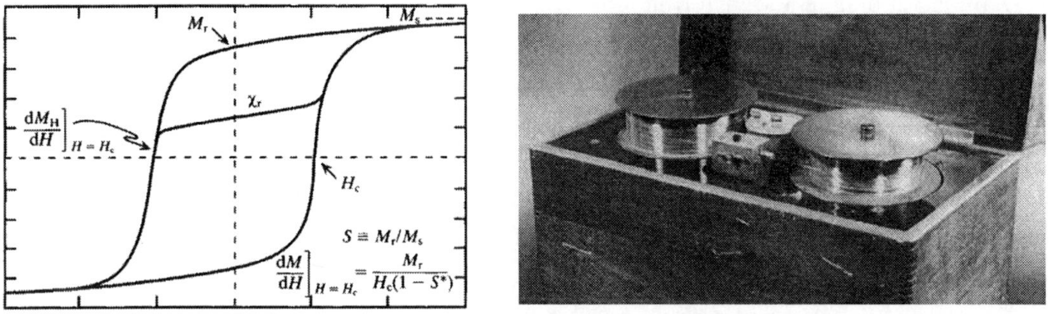

Figure 2. Showing hysteresis loop for ferromagnetic material used on carriers magnetic (χ_r - reversible magnetic susceptibility along a minor loop) [16] and magnetic wire recording (V. Poulsen, 1898)

When magnetizing only via an external magnetic field $\mathbf{M(H_{ext})}$, it is assumed:

- invariability of the chemical composition and microstructure of the material,
- negligible small impact of changes in stress and mechanical strain on the magnetization process of the material,
- negligible small impact of changes in material temperature during the magnetization cycle.

Stress inducted magnetization $\mathbf{M}(\sigma_{ij})$, resulting from accidental impact:

- stress state,
- magneto-mechanical effects,
- internal friction and energy losses,
- changes in chemical composition and microstructure,

has different properties than $\mathbf{M(H_{ext})}$. Especially, when stresses and deformations of the material are generated in the material in the presence of a weak Earth's magnetic field [18-25]. Without the user's knowledge, valuable magnetic information about the history of unknown material strain and progressive material degradation $D = D(t)$ is stored in the ferromagnetic material, which can be used to [2, 23]:

- diagnosing technical condition of DUT,
- monitoring the quality of operation, service and renovation,
- support for the accident investigation process.

Temperature inducted magnetization $\mathbf{M}(\Delta T)$ also has different properties than $\mathbf{M(H_{ext})}$. The magnetic properties of ferromagnetic and domain structure change strongly as the temperature of the material changes [10, 13]. Local and volumetric changes in material temperature during DUT operation are due to:

- the temperature of the surrounding environment (air, water);
- the amount of energy generated in the material, depending on the speed of deformation changes and microstructure parameters (type, grain size, number of inclusions and defects);
- conditions for energy dissipation in the material and energy reception from the material by the surrounding medium;

3.2. Magneto-mechanical damping

Considering five main contributions to the total energy of a ferromagnet without an external field (exchange energy W_{ex}, magnetocrystalline anisotropy energy W_k, magneto-elastic or magnetostrictive energy W_λ, magnetostatic energy W_m, energy of magnetic domain walls W_W), four main mechanisms of magneto-mechanical damping may be defined [14]:

– Magnetoelastic hysteresis damping (${Q_h}^{-1}$)

- Macroeddy-current damping (Q_a^{-1})
- Microeddy-current damping (Q_u^{-1})
- Damping at magnetic transformations (Q_{PhT}^{-1}) e.g., at Curie and N´eel temperatures, spin-flip transitions etc.

Therefore, the total magneto-mechanical damping Q_M^{-1} in ferromagnets can be considered as a sum of these components:

$$Q_M^{-1}(\varepsilon, f, T) = Q_h^{-1}(\varepsilon, f, T) + Q_a^{-1}(f, T) + Q_u^{-1}(f, T) + Q_{PhT}^{-1} \quad (10)$$

where, contrary to Q_a^{-1} and Q_u^{-1}, the hysteretic contribution Q_h^{-1} depends on the strain amplitude ε [14, 23, 24].

3.3. Impact of the DUT hull

To remotely monitor the magnetization status of the DUT internal components, it is necessary to consider the shielding impact of the hull for:
- direct current (static) magnetic field (DC),
- alternating current magnetic field (AC).

Since in contrast to electrical fields no isolated magnetic charges (monopoles) exist, magnetic flux lines are always self-contained; they have no beginning and no end. Consequently, there is no such thing as a magnetic insulator (the principle of superconductivity is excluded here) [24, 25]. While electromagnetic fields with frequencies above approx. 1 kHz can be very well shielded by thin metal foil or meshes made of materials with high electrical conductivity (Faraday cage principle), static or low-frequency electromagnetic fields require more effort. If the frequencies are sufficiently low, the electric and magnetic fields must be considered and shielded independently of each other. The big problem with shielding the low-frequency magnetic field is an advantage for the PMSO. Observation of internal DUT processes can be performed remotely in its vicinity.

The shielding effect of a housing depends on the permeability of its material, the form and size of the housing as well as on the thickness of its walls. Analytical calculation provides a solution for only a few forms. For example, the attenuation of DC magnetic field by DUT hull with single layer is described by a formula (11)

$$A = \frac{H_{out}}{H_{in}} \approx \frac{\mu_m}{4}\left(1 - \frac{RI_1}{RO_1}\right) \quad (11)$$

where A - attenuation @ DC; $\mu_m = \mu_m(\mathbf{H})$ - permeability value of material; RI_1 - inside radius of the single layer; RO_1 - outside radius of the single layer.

Those results can however be used as reference when estimating the shielding effect of other housings. It is important to note that the value for the permeability of a material used for the purposes of a calculation depends on material composition and level of its damage, shield geometry, and field intensity. Because the characteristics $\mu_m(\mathbf{H})$ of ferromagnetic is strongly non-linear in the medium and high magnetic field intensities, this ferromagnetic hull not only weakens the field strength generated by internal DUT elements, but also changes the spectrum of the signal. Near the DUT, magnetizing frequency harmonics will also be present, so the level of total harmonic distortion (THD) is a diagnostic symptom.

A strict mathematical analysis of ferromagnetic shielding of AC magnetic field is also very difficult and so the literature gives useful approximate equations for the shielding effectiveness of common box shapes such cylinders, spheres and cubes [28]. For AC magnetic field the material permeability has a real component μ' and an imaginary component μ'' and the overall permeability is given by:

$$\mu_m = \mu_m' - j \cdot \mu_m'' \tag{12}$$

The Figure 3 shows frequency characteristics.

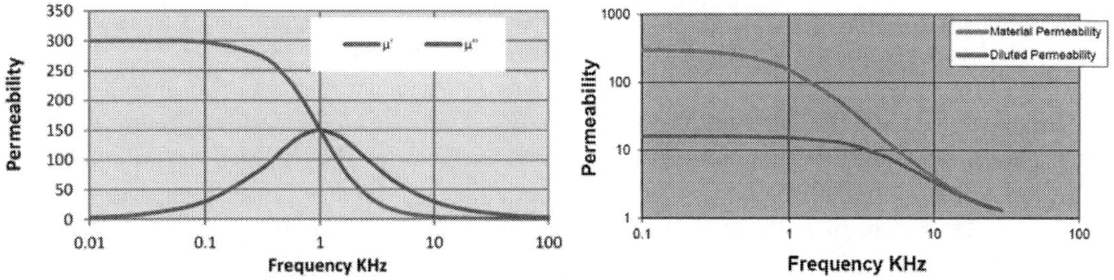

a) b)

Figure 3. Magnetic permeability of ferromagnetic steel with initial (low frequency) permeability $\mu_{m,0}' = 300$, and a relaxation frequency $f_m = 1\ kHz$: a) without influence of diluted component; b) with influence of diluted component [27]

The equations which describes the real component of $\mu_m(f)$ characteristic is:

$$\mu_m' = \frac{(\mu_{m,0}' - 1)}{[1 + (f/f_m)^2]} + 1 \tag{13}$$

At high frequencies this equation is asymptotic to unity, and this is typical of many magnetic materials especially ferrites. Some workers have measured a higher asymptote for steel, and so the above equation needs to be modified to:

$$\mu_m' = \frac{(\mu_{m,0}' - \mu_\infty')}{[1 + (f/f_m)^2]} + \mu_\infty' \tag{14}$$

where μ_∞' is the high frequency permeability. For instance, Bowler measured $\mu_\infty' = 78$ on his steel sample [29].
The loss component shows a resonant characteristic with an amplitude equal to that of the real component at the frequency f_m. Its equation is:

$$\mu_m'' = \frac{\mu_{m,0}'(f/f_m)}{[1 + (f/f_m)^2]} \tag{15}$$

4. Passive magnetic state observer

Having basic theoretical knowledge of magnetism, the research task boils down to:
- selection of the measuring path and its calibration;
- measuring the existing magnetic field near the DUT;
- identification of magnetic anomalies and their sources using measurement data processing algorithms, the basic design features of DUT and the expected signal spectrum;
- determining the current working conditions of the DUT and its health using statistical standards and diagnostic criteria;
- analysis of trends in estimators to predict the technical condition of DUT.

The reliability of diagnosis and prognosis increases as the diagnostician and software learn new capabilities of the diagnostic method.

4.1. Metrological path – capabilities

Magnetic field measurements and analysis are widely used in various applications which results in an increasing market demand for mass-produced magnetometers: magneto-resistive (AMR, GMR, TMR), magneto-induction (MI) and Hall (Hall, Q-Hall) and low their price. Only very high sensitivity (with fT and pT resolution) and low-noise scalar (cesium, rubidium) and vector magnetometers (SQUID, fluxgate) used, among others in geophysics, geology, satellite technology, medicine (magnetoencephalography, magnetocardiography), aerial and marine reconnaissance, detection of unexploded ordnance (UXO), have a price several times higher than other sensors of physical size. The e-compass magnetometers using in smartphones with < 150 nT sensitivity (the Earth magnetic field is about 50000 nT) cost lower than 5 Euro. In the 21st century, there are no technical and technological limitations for the design of applications based on magnetic field measurements from some fT to several T, with divided into subrange or dynamic change in gain. Unfortunately, the dynamic development of the analogue and digital magnetometers market is not reflected in the area of NDT and SHM systems. Only approx. 0.04% of magnetometers produced goes to the NDT and SHM applications, despite the growing share of magnetometers in production quality control systems and car monitoring systems.
The price of mass-produced magnetometers and their datasheets are a reference point to solutions offered by manufacturers of NDT and SHM applications and prototyping.

4.2. Measuring track

Three types of digital magnetometers (MI, TMR, RTD-fluxgate) [30 - 35] with operating data rate (ODR) up to 500 Hz were used in the study. The spectrum of the measured signal contains aliases – weakness of cheap magnetometers (e-compasses), which after considering aliases in data analysis [2] is the strength of PMOS. The limitation of the upper frequency of observation depends on the type of magnetometer and its sensitivity.

4.3. Data analysis (numerical post-processing)

The raw measurement data $B_m(k)$ and its components are described by the relation (16)

$$B_m(k) = \underbrace{B_p(k) + noise(k)}_{Result\ of\ a\ measurement} = \underbrace{B_{p,LPF}(k) + B_{p,BPF}(k) + B_{p,HPF}(k) + noise(k)}_{Analyzed\ signal\ components} \quad (16)$$

where: $B_{p,LPF}(k)$ are low-frequency components (process trends), $B_{p,BPF}(k)$ are band components resulting from DUT design features and observed processes; $B_{p,HPF}(k)$ are high-frequency components; $noise(k)$ - colorful noise whose spectrum contains components: $1/f$, $1/f^2, \dots, 1/f^n$ and white noise.
Only knowledge of the real frequency characteristics of the magnetometer as well as dynamic properties and noise level of the measuring path enables:

- reliable observation of dynamic phenomena occurring on the DUT with frequencies higher than the sampling frequency, which the authors used in the passive magnetic state observer;
- reliable DUT diagnosis based on a typical diagnostic criterion: the detected magnetic anomaly must be greater by min. 2 times (3dB) relative to the noise level.

As the noise level decreases, it increases exponentially:

- likelihood of reliable diagnosis (POD) and DUT status prognosis;

- the distance from which the diagnostic symptoms are detected, among others:
  a) single magnetic anomalies,
  b) groups of cyclical anomalies generated e.g. by gear teeth or compressor and turbine blades
  b) average level of anomaly generated by DUT (in total)
  b) technical condition of DUT ferromagnetic elements (internal and external).

Signal decomposition is implemented numerically using a software prototype and generally known algorithms among others: numerical filtration, FFT analysis, wavelet analysis, demodulation of amplitude modulation (AM) and frequency modulation, 3D analysis of vector components.

5. Results

Examples of PMOS capabilities in the area of EMI, CM, NDT and SHM as well as failure state tests are illustrated in Figure 4 – Figure 9. The obtained research results showed the possibility of extending the functionality of the Metal Magnetic Memory (MMM) method [15, 23] and the need to correct substantive errors in the standard ISO-24497 [36].

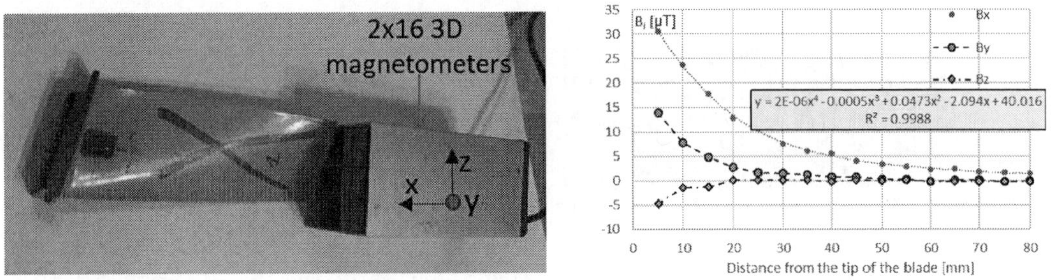

Figure 4. Impact of observation distance on the distribution of the magnetic field near the R1 working blade of a SO-3 engine compressor - a simple test enables the optimal selection of the PMSO measuring range and the type of magnetometer as well as the coefficients of the influence of DUT components on the $\mathbf{B}_p$ signal

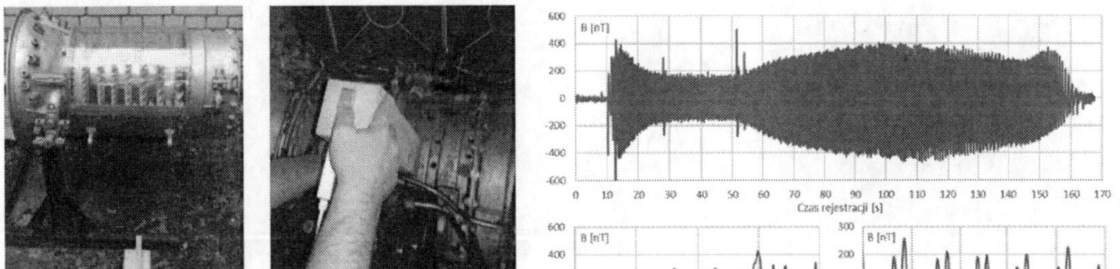

Figure 5. Cold start of SO-3 turbine engine ($n_{max} \approx 2000\ rpm$): remote measurement of instantaneous rotational speed from $0 \longrightarrow n_{max} \longrightarrow 0$ and identification of the technical condition of the middle support by MI PMOS, measuring distance up to 1.0 m from the engine, sensitivity of PMOS about 1 nT, ODR = 250 Hz [37]

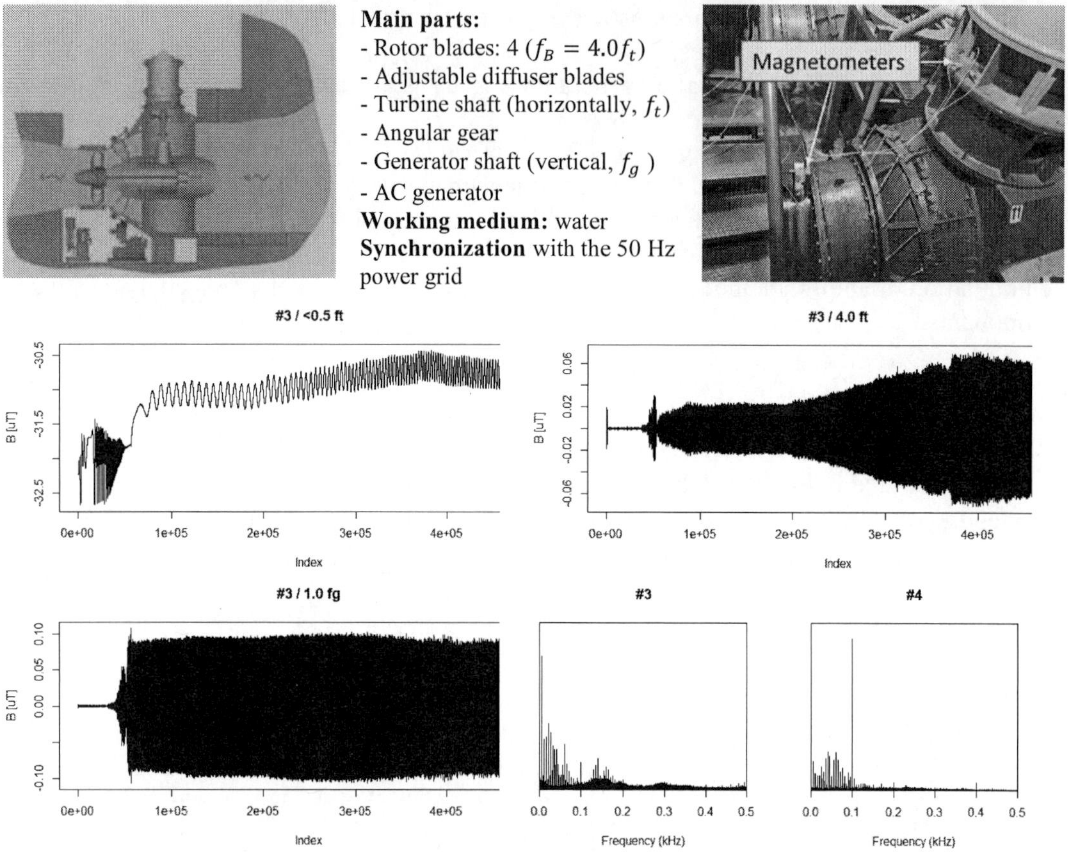

Figure 6. Magnetic field disturbances near the Kaplan turbine - possible remote monitoring of turbine operation parameters and technical condition of its components by PMOS [38]

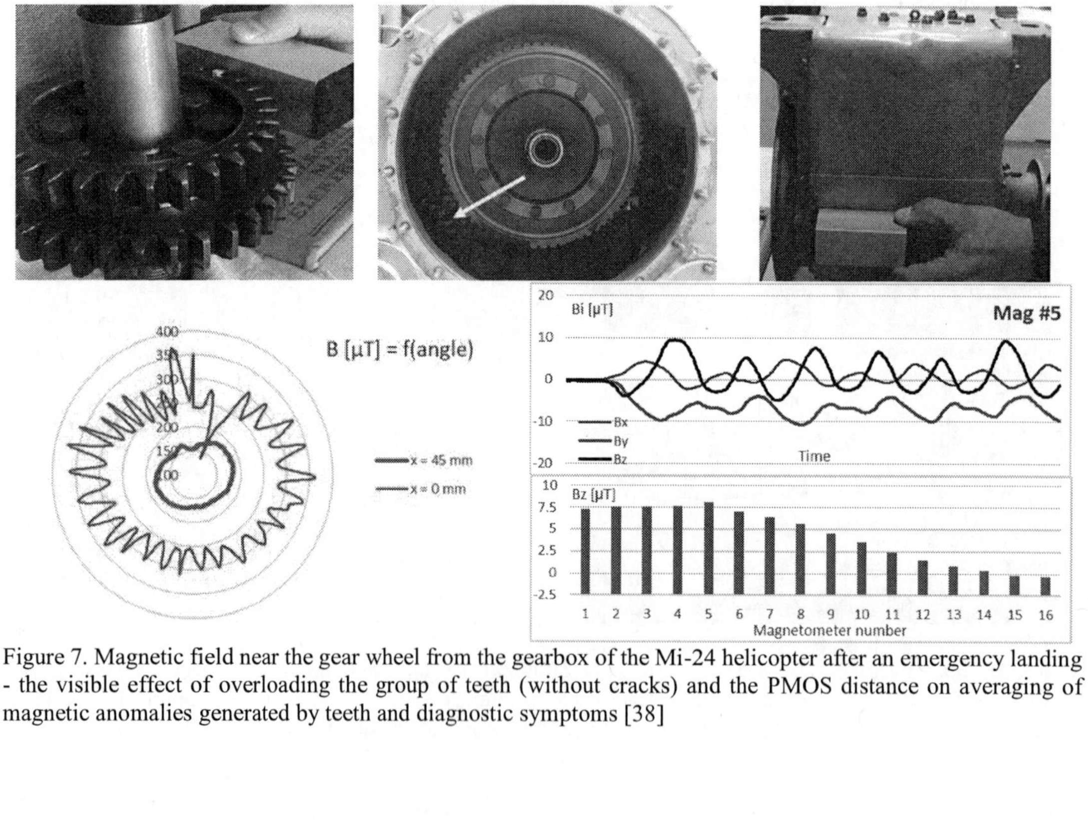

Figure 7. Magnetic field near the gear wheel from the gearbox of the Mi-24 helicopter after an emergency landing - the visible effect of overloading the group of teeth (without cracks) and the PMOS distance on averaging of magnetic anomalies generated by teeth and diagnostic symptoms [38]

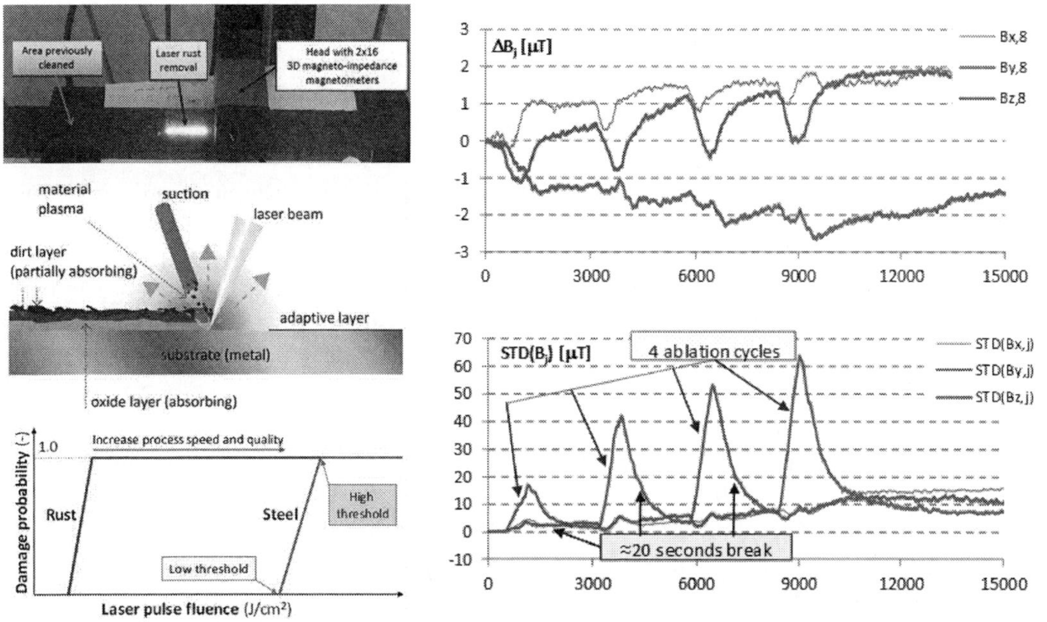

Figure 8. Monitoring of laser clean rust (ablation) by low-cost PMSO [39]

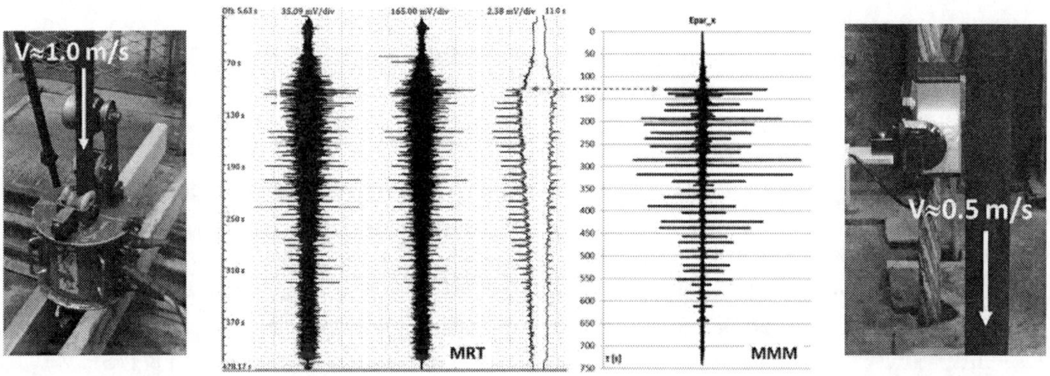

Figure 9. Comparative studies of MRT and PMSO (MMM) methods in the diagnosis of mining ropes – the synthetic aperture of the magnetometer array provides better signal-to-noise ratio (SNR) than the **M**agnetic **R**ope **T**esting (MRT) method - active magnetic state observer (**M**agnetic **F**lux **L**eakage, MFL) using in NDT [40]

Conclusion

The theoretical foundations of the passive magnetic state observer are presented. Possible functions of PMSO have been experimentally verified for:

- non-contact monitoring of electromagnetic interference near the DUT,
- non-contact condition monitoring of DUT,
- nondestructive testing – most reliable then metal magnetic memory method,
- non-contact structural health monitoring,
- failure cause analysis,
- prognostic and health management.

Further research work will be focused on increasing the measurement capabilities up to 25 kHz in the pT range, improving research methodologies and verifying diagnostic symptoms on dynamic objects.

Reference

1. PN-EN 13306: 2018-01 Maintenance – Maintenance Terminology.
2. M. Witos 'Increasing the Durability of Turbine Engines Through Active Diagnostics and Control', Research Works of AFIT, 2011, 29, 1-324, (in Polish), doi: 10.13140/RG.2.1.4341.4560.
3. J. J. Xiong, R. A. Shenoi, 'Fatigue and Fracture Reliability Engineering', Springer - Verlag London 2011, doi: 10.1007/978-0-85729-218-6.
4. B. S. Dhillon, 'Safety, Reliability, Human Factors, and Human Error in Nuclear Power Plants', CRC Press / Taylor & Francis Group, Boca Raton-London-New York, 2017.
5. M. Witos, M. Wachłaczenko, J. Niemczyk, 'A Holisty Approach to Structural Health Monitoring of Turbomachinery', The e-Journal of Nondestructive Testing, 2018, 12, https://www.ndt.net/article/shmndt2018/papers/SHM-NDT-2018_paper_29.pdf.
6. M. Woch, 'Reliability Analysis of the PZL-130 Orlik TC-II Aircraft Structural Component under Real Operating Conditions', Eksplotacja i Niezawodność - Maintenance and Reliability, 2017, 19(2), 287-295, doi: 10.17531/ein.2017.2.17.
7. A. A. Rupp, J. Templin, R. A. Henson, 'Diagnostic Measurement: Theory, Methods, and Applications', The Guilford Press, New York 2010.
8. J. Dobrzański, 'Materiałoznawcza Interpretacja Trwałości Stali dla Energetyki', Tom 2, Open Access Library, International OCSCO World Press, 2016.
9. A. A. Shaniavsky, 'Tolerance Fatigue Failures of Aircraft Components. Synergetics in Engineering Applications', Publishing House of Scientific-and-Technical Literature "Monography", Ufa, 2003 (in Russian).
10. R. E. Newnham, 'Properties of Materials. Anisotropy, Symmetry, Structure', Oxford University Press, Oxford 2004.
11. Torre E. D., 'Magnetic Hysteresis', IEEE Press, Institute of Electrical and Electronics Engineers, Inc. New York 1999.
12. D. J. Craik, R. S. Tebble, 'Ferromagnetism and Ferromagnetic Domains', Tom 4. North-Holland Publishing Company, 1965 (digital version 2011).
13. R. M. Bozorth, 'Ferromagnetism', IEE Press, New York 1993.
14. M. S. Blanter, I.S. Golovin, H. Neuhäuser H.-R. Sinning, 'Internal Friction in Metallic Materials. A Handbook', Springer-Verlag Berlin Heidelberg 2007.
15. https://www.ndt.net/index.php (e-journal & Open Access database of NDT/SHM).
16. H. M. Bertram, 'Theory of Magnetic Recording', Cambridge University Press, 1994.
17. E. F. Gallaudet, 'Relation between Length, Elasticity and Magnetization of Iron and Nickel wires', PhD Dissertation, Johns Hopkins University, Baltimore, Maryland 1896.
18. H. Nagaoka, K. Honda, 'Researches on Magnetostriction', The Journal of the College of Science, Imperial University of Tokyo, Section 353, 1898-1925.
19. C.W. Barrows, 'Correlation of the Magnetic and Mechanical Properties of Steel'. Scientific Papers of the Bureau of Standards, No 272, Government Printing Office, Washington 1916.
20. J. R. Hobbie, Jr., 'Magnetostriction with Small Magnetic Fields', Physical Review, 1922, Vol. XIX, No 5.
21. I. M. Robertson, 'Magneto-Elastic Behaviour of Steel for Naval Applications', MRL Technical Report, MRL-TR-90-27, DTSO Material Research Laboratory, 1991.
22. M. J. Dapino, 'On Magnetostriction Materials and their Use in Adaptive Structures', Structural Engineering and Mechanics, 2004, 17(3-4).
23. V. T. Vlasov, A. A. Dubov, 'Physical Bases of the Metal Magnetic Memory Method', ZAO "Tisso" Publishing House, Moscow 2004.

24. V. T. Vlasov, A. A. Dubov, 'Physical Theory of the "Strain – Failure Process', Publishing House "SPEKTR", Moscow, 2013.
25. M. Witos, 'The Reference Signal of Geomagnetic Field for MMM Expert System', Key Engineering Materials, 2012, 518, 384-395, doi: 10.4028/www.scientific.net/KEM.518.384.
26. S. Celozzi, R. Araneo, G. Lovat, 'Electromagnet Shielding', IEEE Press / A John Wiley & Sons, Inc., New Jersey 2008, published simultaneously in Canada.
27. A. Payne, 'Shielding of Magnetic Fields by Eddy Currents', 2016, e-mail: paynealpayne@aol.com
28. SEKELS BROCHURE: http://www.sekels.de/fileadmin/PDF/Englisch/06_Brochure_Magnetic_Shielding.pdf.
29. N. Bowler, 'Frequency-dependence of Relative Permeability of Steel'. CP820 Review of Quantitive Nondestructive Evaluation 2006, Vol 25. http://home.eng.iastate.edu/~nbowler/pdf%20final%20versions/conferences/QNDE2005Bowler.pdf
30. P. Ripka (ed.), 'Magnetic Sensor and Magnetometers', Artech House, Boston 2000.
31. W.Gőpel, J. Hesse, J. N. Zemél (eds), 'Sensors. A Comprehensive Survey. Vol. 5 Magnetic sensor', OVCH Verlagsgesellschaft mbH,1989.
32. R. Varga, P. Klein, R. Sabol, et al., Magnetically Bistable Microwires: Properties and Application for Magnetic Field, Temperature, and Stress Sensing, [in:] Zhukov A.(ed.) 'High Performance Soft Magnetic Materials', Springer, 2017, doi: 10.1007/978-3-319-49707-5.
33. https: //www.aichi-mi.com/e-home-new/.
34. https://www.nxp.com.
35 https://www.micromagnetics.com.
36. ISO-24497:2007 Non-destructive testing — Metal magnetic memory. Part 1: Vocabulary. Part 2: General requirements. Part 3: Inspection of welded joints
37. P. Mazurek, M. Witos, J. Kwaśniewski, M. Roskosz, 'The use of Magneto-mechanical Effects in the Diagnosis of SO-3 Turbine Jet Engine' (in Polish), 1st Conference 'Physics in Aviation', September 16-18, 2019, Dęblin.
38. M. Witos, M. Oravec, M. Šmelko, P. Lipovsky, K. Bogucki, 'Passive Magnetic Observer' (in Polish), 1st Conference 'Physics in Aviation', September 16-18, 2019, Dęblin.
39. W. Napadłek, M. Witoś, M. Żokowski, 'Influence of Parameters of Laser Ablation on Magnetization Ferromagnetics', 17th Czech and Slovak Conference on Magnetism CSMAG'19, June 3 –7, 2019, Košice, Slovakia, doi: 10.13140/RG.2.2.11394.12483 (CSMAG19 P9-17 poster).
40. M. Witos, M. Żokowski, J. Kwaśniewski, 'Selected Aspects of MRT and MMM Research in Mining Industry', XIV International Conference 'Safety Operation of Mining Transport Equipment', November 7-9, 2018, Wisła, Poland, [in:] A. Tytko, M. Wójcik (eds), 'Nowe trendy i doświadczenia w eksploatacji urządzeń transportowych', Wydawnictwo AGH, Kraków 2019 (in Polish), [in:] doi:10.13140/RG.2.2.24233.65128 (presentation in English).

AUTHOR INDEX